Lecture Notes in Mathematics

A collection of informal reports and seminars
Edited by A. Dold, Heidelberg and B. Eckmann, Zürich

182

Leonard D. Baumert
California Institute of Technology
Pasadena, CA / USA

Cyclic Difference Sets

Springer-Verlag
Berlin · Heidelberg · New York 1971

ISBN 3-540-05368-9 Springer-Verlag Berlin · Heidelberg · New York
ISBN 0-387-05368-9 Springer-Verlag New York · Heidelberg · Berlin

Offsetdruck: Julius Beltz, Weinheim/Bergstr.

CYCLIC DIFFERENCE SETS

A fairly comprehensive survey of the general theory of cyclic difference sets is given below. The aim of this survey is to provide a cohesive presentation of the known facts as well as an introduction to some of the outstanding problems. The more general topics of block designs and difference sets in finite groups are introduced but only those aspects of these subjects which shed some light on problems arising for cyclic difference sets are developed.

It is not expected that many will wish to read this survey sequentially from the beginning. For this reason the chapters and to a lesser degree the sections within them are largely independent of each other, having been written that way in order to encourage the reader to skip around and follow his own interests. However a certain familiarity with the contents of Chapter I is presupposed elsewhere. Beyond this, interconnections between the various sections and chapters are indicated when they seem relevant. This structure, coupled with the aim of making the later material understandable to as many as possible, has led to the anomaly that, in some cases, quite elementary concepts are defined in the later chapters, whereas these same concepts, and a great deal more, were presupposed in earlier sections.

In addition to the specific references inserted in the text, the books of Marshall Hall, Jr., "Combinatorial Theory", Blaisdell Publishing Company, 1967, of H. B. Mann, "Addition Theorems", Interscience Publishers, 1965, and of H. J. Ryser, "Combinatorial Mathematics", Carus Mathematical Monograph No. 14, 1963, may be used as general references for a large part of this material.

This survey was compiled in connection with research carried out at the Jet Propulsion Laboratory, California Institute of Technology, under Contract No. NAS 7-100, sponsored by the National Aeronautics and Space Administration.

CONTENTS

I. INTRODUCTION

The main purpose of this chapter is to provide the basic definitions and vocabulary of the study of difference sets so that subsequent chapters need not be interrupted at inopportune times by the introduction of such material. Thus none of the concepts are pursued in detail - such development being deferred to the appropriate part of a later chapter.

A. Difference Sets

A <u>v, k, λ - difference set</u> $D = \{d_1, \ldots, d_k\}$ is a collection of k residues modulo v, such that for <u>any</u> residue $\alpha \not\equiv 0 \pmod v$ the congruence

$$d_i - d_j \equiv \alpha \pmod v \tag{1.1}$$

has exactly λ solution pairs (d_i, d_j) with d_i and d_j in D. [The terms difference set, cyclic difference set and v, k, λ - difference set are interchangeable below; the later two being used when there is some reason to stress either the contrast with general group difference sets or the particular <u>parameters</u> v, k, λ involved.]

As an immediate consequence of this definition one has that the relation

$$k(k - 1) = \lambda(v - 1) \tag{1.2}$$

necessarily holds among the parameters v, k, λ.

Given any positive integer v there are certain obvious difference sets modulo v. These are:

(i) the null set $D = \emptyset$

(ii) all singletons $D = \{i\}$, $0 \leq i \leq v - 1$

(iii) $D = \{0, 1, \ldots, v - 1\}$

(iv) $D = \{0, 1, \ldots, i - 1, i + 1, \ldots, v - 1\}$ $\quad 0 \leq i \leq v - 1$.

These difference sets are called trivial and are quite often either ignored or treated only as limiting cases. Note that the parameters for these difference sets are $v, k, \lambda = v, 0, 0$; $v, 1, 0$; v, v, v and $v, v-1, v-2$ respectively. If one introduces the additional parameter $n = k - \lambda$, equation (1.2) above shows that these trivial difference sets arise if and only if $n = 0$ or 1. Hence the assumption $n \geq 2$, which is often made implicitly, serves to exclude all the trivial difference sets.

Some non-trivial difference sets are:

$$D_1 = \{1, 2, 4\} \mod 7,$$
$$D_2 = \{0, 3, 5, 6\} \mod 7,$$
$$D_3 = \{0, 1, 3, 9\} \mod 13,$$
$$D_4 = \{1, 4, 5, 6, 7, 9, 11, 16, 17\} \mod 19.$$

Here the parameters v, k, λ are $7, 3, 1$; $7, 4, 2$; $13, 4, 1$ and $19, 9, 4$ respectively.

B. Shifts, Equivalence, Complements

Given the difference set $D = \{d_1, \ldots, d_k\} \mod v$, then for any integer s the set $\{d_1 + s, \ldots, d_k + s\} \equiv D + s$ taken modulo v is also a difference set; it is called a shift (or cyclic shift) of the original set D. Again, given any integer t, prime to v, the set $\{td_1, \ldots, td_k\} \equiv tD$ taken modulo v is also a difference set with the same parameters v, k, λ. If D_i and D_j are two difference sets having the same parameters and if $D_i \equiv tD_j + s \pmod{v}$ for any integers t, s with t prime to v, then D_i and D_j are said to be equivalent difference sets. For any particular set of parameters v, k, λ satisfying equation (1.2) there may be no difference sets, a single difference set (up to equivalence) or several inequivalent difference sets having these parameters. In fact, Gordon, Mills and Welch (1962) have shown that given any $\ell \geq 0$, there exist values of v, k, λ such that there are at least ℓ pairwise inequivalent difference sets with these parameters. (This result is discussed in V. A. below).

Among the examples given in I. A. above note that D_1 and D_2 are related, that is

$$D_1 + D_2 = \{0,1,\ldots,v - 1\}.$$

For this reason D_1 and D_2 are said to be <u>complementary</u> difference sets. If D is a difference set with parameters v, k, λ its complement D^* is a difference set with parameters $v^*, k^*, \lambda^* = v, v - k, v - 2k + \lambda$. Recalling that $n = k - \lambda$ one sees that both v and n are invariant under complementation of the difference set. For most purposes it is sufficient to consider only one of a pair of complementary difference sets. This is frequently done by insisting that k be less than $v/2$. (Equation (1.2) shows that $k = v/2$ does not occur).

As will become abundantly clear, the parameters v, n are the most fundamental of the four - v, k, λ, n. For this reason it is sometimes useful to express k, λ in terms of them. Using (1.2) and $n = k - \lambda$ it follows that

$$k \cdot k^* = k(v - k) = n(v - 1) \quad \text{and} \quad \lambda \cdot \lambda^* = \lambda(v - 2n - \lambda) = n(n - 1).$$

Since $\lambda + \lambda^* = v - 2n$ and $\lambda \geq 1$ for non-trivial difference sets, it follows that

$$\left(\frac{v - 2n}{2}\right)^2 \geq \lambda \cdot \lambda^* = n(n - 1) \geq v - 2n - 1$$

and so

$$n^2 + n + 1 \geq v \geq 4n - 1.$$

The two extreme cases $v = 4n - 1$ and $v = n^2 + n + 1$ correspond to difference sets of the Hadamard type and to the difference sets associated with certain finite projective planes (so-called planar difference sets). Both these types are discussed more fully in Chapter IV - Difference Sets of Special Type.

C. Block Designs, Incidence Matrices, The Incidence Equation

An arrangement of v objects into b sets (called blocks) such that:

(i) each block contains exactly k different objects

(ii) each object occurs in exactly r different blocks

(iii) any pair of objects occurs in exactly λ different blocks

is called a balanced incomplete block design. Evidently the relations

$$bk = vr \tag{1.3}$$

$$r(k - 1) = \lambda(v - 1) \tag{1.4}$$

are satisfied by the parameters b, v, r, k, λ. (The last follows by counting the pairs involving any one fixed object.) If the number of blocks b equals the number of objects v the design is said to be symmetric. In a symmetric design equation (1.3) shows that $k = r$. Thus for symmetric designs equation (1.4) takes the same form as equation (1.2).

Every difference set gives rise to a symmetric block design whose objects are $0,1,\ldots,v - 1$ and whose blocks are $D, D + 1,\ldots,D + (v - 1)$. Note that any cyclic relabeling of the objects of such a design permutes the blocks cyclically in a cycle of length v. Such a block design is called cyclic. Thus every difference set corresponds to a cyclic symmetric balanced incomplete block design. For example, the difference set D_1 of I.A. above corresponds to the block design:

$$\begin{aligned}
B_1 &= D_1 &&\equiv \{1, 2, 4\}\\
B_2 &= D_1 + 1 &&\equiv \{2, 3, 5\}\\
B_3 &= D_1 + 2 &&\equiv \{3, 4, 6\}\\
B_4 &= D_1 + 3 &&\equiv \{0, 4, 5\} \qquad \pmod 7\\
B_5 &= D_1 + 4 &&\equiv \{1, 5, 6\}\\
B_6 &= D_1 + 5 &&\equiv \{0, 2, 6\}\\
B_7 &= D_1 + 6 &&\equiv \{0, 1, 3\}
\end{aligned}$$

Associated with any balanced incomplete block design is a $b \times v$ matrix A of zeros and ones called the incidence matrix of the design. It is constructed by putting $a_{ij} = 1$ if the jth object appears in the ith block and $a_{ij} = 0$ otherwise. For the example above [recalling that the ith block is $D + (i - 1)$ and that the jth object is $j - 1$] the incidence matrix is

$$A = \begin{bmatrix} 0 & 1 & 1 & 0 & 1 & 0 & 0 \\ 0 & 0 & 1 & 1 & 0 & 1 & 0 \\ 0 & 0 & 0 & 1 & 1 & 0 & 1 \\ 1 & 0 & 0 & 0 & 1 & 1 & 0 \\ 0 & 1 & 0 & 0 & 0 & 1 & 1 \\ 1 & 0 & 1 & 0 & 0 & 0 & 1 \\ 1 & 1 & 0 & 1 & 0 & 0 & 0 \end{bmatrix}$$

From the block design definition above it is clear that the associated incidence matrix satisfies the so-called incidence equation

$$A^T A = (r - \lambda)I + \lambda J \tag{1.5}$$

where I is the identity of order v and J is the matrix of all ones of that same order. From this it follows that the determinant of $A^T A$ is given by

$$|A^T A| = [r + (v - 1)\lambda]\,(r - \lambda)^{v-1} . \tag{1.6}$$

If the block design is symmetric, this becomes (using equation (1.4))

$$|A^T A| = k^2 (k - \lambda)^{v-1} \tag{1.7}$$

from which it follows that the incidence matrix A of a non-trivial symmetric balanced incomplete block design is non-singular. Using this fact it can be shown (Ryser, 1950) that the incidence matrix of a symmetric block design is normal. That is, that

$$A^T A = AA^T = (k - \lambda)I + \lambda J \tag{1.8}$$

for symmetric designs.

Since $|A^T A| = |A|^2$ one can conclude from equation (1.8) or (1.7) that a symmetric block design (or a difference set) can only exist for even v if $n = k - \lambda$ is a square. The study of equation (1.8) has led to other significant existence criterions for symmetric block designs and hence for difference sets. Since Chapter II is devoted to difference set existence questions these results are deferred until then.

D. The Characteristic Function and its Autocorrelation Function

Corresponding to the difference set D is the binary sequence $\{a_i\}$ $(i = 0,\ldots,v - 1)$ given by $a_i = 1$ if i is a member of D and $a_i = 0$ otherwise. This is called the characteristic function or incidence vector of the difference set D. [Of course it appeared in I. C. as the first row of the incidence matrix associated with the difference set.] Considered as a binary sequence, it is quite natural to inquire about its autocorrelation function

$$R_a(j) = \sum_{i=0}^{v-1} a_i \, a_{i+j} \qquad (i + j \text{ taken modulo } v)$$

and normalized autocorrelation function

$$\rho_a(j) = \frac{1}{v} R_a(j) .$$

Since $\{a_i\}$ is the characteristic function of a difference set, the autocorrelation function is particularly simple, i.e.

$$R_a(j) = \begin{cases} k & \text{if } j \equiv 0 \text{ modulo } v \\ \lambda & \text{otherwise} \end{cases}$$

Frequently the binary sequence $\{b_i\}_0^{v-1}$ where $b_i = 2a_i - 1$ is considered instead of the characteristic function (note the transformation merely replaces the zeros of $\{a_i\}$ with minus ones). This sequence $\{b_i\}$ has autocorrelation function

$$R_b(j) = \begin{cases} v & \text{if } j \equiv 0 \text{ modulo } v \\ v - 4(k - \lambda) & \text{otherwise} \end{cases}$$

Autocorrelation functions like these are said to be two-level and binary sequences which possess them have found extensive application in digital communications. [See Golomb et al, 1964 for some of these]. Of course, it follows immediately from the definitions that the only binary sequences which have two-level autocorrelation functions are those associated with cyclic differences sets.

E. Multipliers

If t is prime to v and if tD is some shift $D + s$ of the original difference set D, then t is called a multiplier of D. In terms of the associated block design of the difference set (see I.C. above) the mapping $x \to tx$ (mod v) is an automorphism. That is, if A is the incidence matrix of the associated block design, there exists permutation matrices P, Q determined by t of order v such that

$$PAQ = A . \tag{1.9}$$

All known v, k, λ - difference sets have non-trivial multipliers (i.e., multipliers $t \not\equiv 1$ modulo v). The question as to whether this must be so is open. The collection of multipliers of a given difference set forms a group called the multiplier group of that difference set. One of the most useful results in the theory of difference sets is the so-called "multiplier" theorem. This theorem guarantees the existence of multipliers under certain circumstances.

<u>Theorem 1.1.</u> (Hall and Ryser, 1951). If p is a prime dividing n, if p is prime to v and if $p > \lambda$, then p is a multiplier of all difference sets with these parameters v, k, λ.

Chapter III discusses this result and its generalizations. For the present, merely note that for all <u>known</u> non-trivial difference sets the condition $p > \lambda$ is superfluous. Again since (another open question) all <u>known</u> non-trivial cyclic difference sets have $(n,v) = 1$ one has that for all <u>known</u> non-trivial cyclic difference sets every divisor t of n is a multiplier.

F. <u>The Hall-polynomial, w-multipliers</u>

Instead of the difference set itself, it is often convenient to deal with the polynomial

$$\theta(x) = x^{d_1} + \dots + x^{d_k} . \tag{1.10}$$

This polynomial has been called the <u>Hall-polynomial</u> of the difference set, the <u>generating polynomial</u> of the difference set or the <u>difference set polynomial</u>.

In terms of this polynomial the difference set property is

$$\theta(x)\theta(x^{-1}) = \sum_{i,j}^{k} x^{d_i - d_j} \equiv n + \lambda(1 + x + \cdots + x^{v-1}) \qquad (\text{mod } x^v - 1) . \tag{1.11}$$

If $\zeta_v \neq 1$ is any vth root of unity this congruence yields

$$\theta(\zeta_v)\, \theta(\zeta_v^{-1}) = n . \tag{1.12}$$

This shows that a non-trivial difference set is intimately connected with the factorization of n in the field of vth roots of unity.

If t is a multiplier of the difference set D then

$$\theta(x^t) \equiv x^s\theta(x) \pmod{x^v - 1}. \tag{1.13}$$

More generally, if w divides v, define a w-multiplier to be any integer t, prime to w, for which there exists an integer s satisfying

$$\theta(x^t) \equiv x^s\theta(x) \pmod{x^w - 1}. \tag{1.14}$$

Clearly s may be assumed to be non-negative. Further, a w-multiplier is a w_1-multiplier for all divisors w_1 of w. Thus, an ordinary multiplier is a w-multiplier for every divisor w of v.

On occasion it is possible to demonstrate that a hypothetical difference set must have a w-multiplier even when it is not possible to show the existence of a multiplier. This often leads to non-existence proofs which can not be deduced from the strict multiplier theory.

G. Group Difference Sets

A difference set in a group G of order v is a set $\{g_1, \ldots, g_k\}$ of distinct elements of G such that the equation

$$g_i g_j^{-1} = g$$

has exactly λ solutions for every g in G, $g \neq 1$.

Group difference sets are called non-Abelian, Abelian, or cyclic according to whether the group is non-Abelian, Abelian or cyclic. The difference sets considered above (I.A., etc.) correspond to group difference sets for cyclic groups (i.e., they are cyclic difference sets under this terminology). Thus group difference sets constitute a generalization (due to Bruck, 1955) of those previously under consideration. Every group difference set gives rise to a symmetric block design in much the same manner as demonstrated for cyclic difference sets in I.C. above. But not every symmetric block design corresponds to a difference set in some group G. [For example, the $v, k, \lambda = 31, 10, 3$ design (listed in Hall, 1967, p. 293) could only correspond to a cyclic difference set since the only group of order 31 is cyclic. But this design is obviously not cyclic.] Thus group difference sets occupy a truly middle ground between symmetric block designs and the cyclic difference sets of concern here.

The main reason for introducing general group difference sets into this discussion at all is that some of the major outstanding problems are only of concern for cyclic difference sets. Thus, in subsequent chapters a few facts about general group difference sets are mentioned in connection with these problems. The purpose being to point out that the difficulties arise only because of the cyclic nature of things and thus cannot be resolved solely by techniques which apply more generally.

An example of a group difference set, which is not cyclic, is the set $D = \{a, b, c, d, ab, cd\}$ in the Abelian group of order 16 generated by a, b, c, d, where $a^2 = b^2 = c^2 = d^2 = 1$. This set has parameters $v, k, \lambda = 16, 6, 2$.

II. EXISTENCE QUESTIONS

The main questions regarding difference sets are: When do they exist? How do you construct them? How many (inequivalent ones) are there? Even though there is considerable overlap between the areas defined by these questions, there seems to be some value in treating them separately. Thus, this chapter is primarily concerned with conditions necessary for the existence of difference sets. For the most part only number theoretic results involving the parameters v, k, λ, n and their divisors are considered. Some existence tests of a more constructive nature are presented in Chapter III - Multipliers and Constructive Existence Tests.

A. The Main Existence Problems

As the title of this chapter implies, the existence question for difference sets is unsolved. That is, given parameters v, k, λ it is (in general) impossible to decide (short of an exhaustive search) whether or not a difference set with these parameters exists. Nevertheless, significant progress has been made. Perhaps the most important test is the obvious relationship

$$k(k-1) = \lambda(v-1) \text{ or } k^2 = \lambda v + n . \qquad (2.1)$$

A sub-problem of this general existence question is the curious fact that no difference sets are known which have $(v,n) > 1$ [or equivalently $(k,v) > 1$], though no proof of this has been given. Here one must be careful to distinguish between cyclic difference sets and general group difference sets. For, there do exist group difference sets with $(v,n) > 1$; the example given in I.G. above has parameters $v, k, \lambda, n = 16, 6, 2, 4$. Thus, if there are no cyclic difference sets with $(v,n) > 1$, the proof must be intrinsically cyclic.

Another outstanding existence problem arises when one notes that there exists an infinite number of difference sets with $\lambda = 1$. Specifically $v, k, \lambda = p^{2j} + p^j + 1, p^j + 1, 1$ for all primes p (see Section V.A. for details of their construction). It has been conjectured that for every $\lambda \geq 2$, there exists only a finite number of difference sets. This conjecture is wide open; it has not been either proved or refuted for any single value of λ. Of course the same conjecture can be made for symmetric block designs and again the problem is open.

B. The Bruck-Ryser-Chowla Theorem

As pointed out in Section I.C. above, the Incidence Equation, i.e.,

$$A^T A = nI + \lambda J \tag{2.2}$$

holds for all symmetric block designs. (Here A is the $v \times v$ incidence matrix of the design, J is the $v \times v$ matrix of all ones and I is the identity of order v). Associate a linear form with each row of the incidence matrix A, according to the rule

$$L_i(x) = \sum_{j=1}^{v} a_{ij} x_j$$

where $x = (x_1, \ldots, x_v)$ is a vector of indeterminates x_j. Then equation (2.2) takes the form

$$L_1^2(x) + \cdots + L_v^2(x) = n(x_1^2 + \cdots + x_v^2) + \lambda(x_1 + \cdots + x_v)^2 \qquad (2.3)$$

The study of these equations ((2.2) and (2.3)) has produced a number of existence criterions for symmetric block designs as well as for certain more specialized configurations.

Let $B = nI + \lambda J$ and write A^TA as A^TIA. Then (using the language of quadratic forms) equation (2.2) shows that if a block design exists, then the identity matrix I represents B with a 0, 1 transformation matrix A. Unfortunately, the theory of such matrix representations is not fully developed even when A is allowed to have arbitrary integer coefficients. However, if A is permited rational coefficients, the Hasse-Minkowski theory of rational equivalence of quadratic forms [see Jones (1950) for an exposition of this theory] provides necessary and sufficient conditions for the existence of such a transformation A. Specifically

<u>Theorem 2.1</u> (Bruck-Ryser-Chowla) A $v \times v$ rational matrix A satisfying $A^TA = nI + \lambda J$, when $k(k - 1) = \lambda(v - 1)$, exists if and only if

(i) for v even, n is a square

(ii) for v odd, the equation $z^2 = nx^2 + (-1)^{(v-1)/2}\lambda y^2$ has a solution in integers x, y, z not all zero.

Thus Theorem 2.1 provides necessary conditions for the existence of symmetric block designs and hence for difference sets. In fact, there is no parameter set v, k, λ satisfying Theorem 2.1 for which it is <u>known</u> that no symmetric block design exists. That is, conceivably the conditions of Theorem 2.1 are sufficient not only for the existence of a rational matrix A but also a 0, 1 matrix. If one restricts attention to cyclic difference sets however, this is no longer the case. For [as is shown later, section II.E] there is no cyclic difference set with parameters $v, k, \lambda = 16, 6, 2$ even though these parameters do satisfy Theorem 2.1, as they clearly must since an example of such a block design can be derived from the non-cyclic difference set given in section I.G. above.

It should be pointed out that Legendre [see Nagell (1951) Theorem 113] provided a simple effective test for the solvability of diophantine equations of the type appearing in (ii) above (see Note 2 below for a statement of this test). Thus criterion (ii) is an effective criterion even when a solution of the diophantine equation is not obvious. As an example of a parameter set excluded by this theorem consider $v, k, \lambda = 43, 7, 1$; this leads to the diophantine equation $z^2 = 6x^2 - y^2$.

Necessity of the conditions of Theorem 2.1 can be proved without recourse to the Hasse-Minkowski theory (see below). Even sufficiency is available for the case $\lambda = 1$ [Hall (1967), p.111] and also whenever n is a square. In particular, when n is a square the rational matrix

$$A = \sqrt{n}\, I + \frac{k - \sqrt{n}}{v} J$$

satisfies equation (2.2).

Proof of Theorem 2.1 [necessity only, Chowla and Ryser (1950)]. From equation (2.2) it follows that $(\det A)^2 = k^2(k - \lambda)^{v-1}$; thus when v is even, $n = k - \lambda$ is a square. For odd v, the number-theoretic result that every positive integer is representable as a sum of four integral squares [see, for example, Nagell (1951)] and Euler's identity

$$(b_1^2 + b_2^2 + b_3^2 + b_4^2)(x_1^2 + x_2^2 + x_3^2 + x_4^2) = y_1^2 + y_2^2 + y_3^2 + y_4^2 \qquad (2.4)$$

where

$$\begin{aligned} y_1 &= b_1x_1 - b_2x_2 - b_3x_3 - b_4x_4 \\ y_2 &= b_2x_1 + b_1x_2 - b_4x_3 + b_3x_4 \\ y_3 &= b_3x_1 + b_4x_2 + b_1x_3 - b_2x_4 \\ y_4 &= b_4x_1 - b_3x_2 + b_2x_3 + b_1x_4 \end{aligned} \tag{2.5}$$

are required. With $n = b_1^2 + b_2^2 + b_3^2 + b_4^2$ (b_i integers) the determinant of the system of equations (2.5) is n^2. Thus Cramer's rule shows that the system may be solved for the x_i's as linear combinations of the y_i's with rational coefficients, the denominators of which are n^2.

When $v \equiv 1 \pmod 4$, the relation $n = b_1^2 + b_2^2 + b_3^2 + b_4^2$ together with $x_v = y_v$ and

$$n(x_i^2 + x_{i+1}^2 + x_{i+2}^2 + x_{i+3}^2) = y_i^2 + y_{i+1}^2 + y_{i+2}^2 + y_{i+3}^2 \tag{2.6}$$

for $i = 1,5,\dots,v-4$ can be used to transform equation (2.3) into an identity in the independent indeterminants $y_1,\dots,y_v$ given by

$$L_1^2(y) + \cdots + L_v^2(y) = y_1^2 + \cdots + y_{v-1}^2 + ny_v + \lambda w^2 \tag{2.7}$$

where $L_1(y),\dots,L_v(y)$ and $w = x_1 + x_2 + \cdots + x_v$ are rational linear forms in $y_1,\dots,y_v$. Since (2.7) is an identity in the y_i's it is valid for all values of y_1, in particular for the value

$$y_1 = \begin{cases} \dfrac{c_2y_2 + \cdots + c_vy_v}{1 - c_1} & \text{for } c_1 \neq 1 \\[2ex] \dfrac{c_2y_2 + \cdots + c_vy_v}{-1 - c_1} & \text{for } c_1 = 1 \end{cases} \tag{2.8}$$

where

$$L_1(y) = \sum_{j=1}^{n} c_j y_j .$$

For this value of y_1 however, $L_1^2(y) = y_1^2$; thus equation (2.7) reduces to an identity in the variables $y_2, \ldots, y_v$. Proceeding in this manner with $y_2, \ldots, y_{v-1}$ in turn, yields the identity

$$L_v^2(y) = ny_v^2 + \lambda w^2$$

where $L_v(y)$ and w are rational multiples of y_v. Now let y_v be a non-zero integer x, which is a multiple of the denominators appearing in L_v and w, then in integers x, y, z $(x \neq 0)$ the equation

$$z^2 = nx^2 + \lambda y^2 \tag{2.9}$$

has a solution.

When $v \equiv 3 \pmod 4$, add nx_{v+1}^2 to both sides of equation (2.3) where x_{v+1} is a new indeterminate. Proceeding as before yields the identity

$$L_1^2(y) + \cdots + L_v^2(y) + nx_{v+1}^2 = y_1^2 + \cdots + y_{v+1}^2 + \lambda w^2$$

where $L_1(y), \ldots, L_v(y)$, x_{v+1} and w are rational linear forms in $y_1, \ldots, y_{v+1}$. Again choosing $y_1, \ldots, y_v$ judiciously implies the identity

$$nx_{v+1}^2 = y_{v+1}^2 + \lambda w^2$$

where x_{v+1} and w are rational multiples of y_{v+1}. Taking y_{v+1} to be a non-zero integer z which is a multiple of the denominators of x_{v+1} and w yields a solution in integers x, y, z $(z \neq 0)$ of the equation

$$nx^2 = z^2 + \lambda y^2 .$$

Combining this equation with that of (2.9) completes the proof of the necessity of the conditions of Theorem 2.1.

Condition (i) of Theorem 2.1 was derived independently by Schützenberger (1949) and by Chowla and Ryser (1950). Condition (ii) was first established, for $\lambda = 1$ only, by Bruck and Ryser (1949) and then generalized to arbitrary λ independently by Chowla and Ryser (1950) and Shrikhande (1950).

As pointed out in section I.C. the incidence matrix A of a symmetric block design not only satisfies equation (2.2) above but also must be normal (Ryser, 1950). That is, it must satisfy

$$AA^T = A^TA = (k - \lambda)I + \lambda J . \qquad (2.10)$$

Thus, when Theorem 2.1 was first established, there was some reason to hope that adding the normality condition would further restrict the possible parameter sets v, k, λ. This was shown not to be true by Albert (1953) for $\lambda = 1$ and by Hall and Ryser (1954) for general λ. That is, the conditions of Theorem 2.1 are sufficient to guarantee the existence of a normal rational matrix A satisfying equation (2.2). Moreover, this solution A also satisfies the condition $AJ = kJ$ which is trivially necessary for block designs.

Hall and Ryser also showed that any set of initial 0, 1 rows, consistent with equation (2.10) above, can be completed to a normal rational matrix A satisfying that equation. [See Hall (1967) p. 275 for the proof of a somewhat more general result.] Clearly, specifying an initial set of 0,1 columns subject to equation (2.10) yields the same result. E. C. Johnsen (1965) has generalized this to the case where r rows and s columns are given, subject to the provisions of equation (2.10) and those imposed by $AJ = JA = kJ$, which are clearly necessary for symmetric block designs. He showed that even here there always exists a normal rational matrix A satisfying equation (2.10) as well as $AJ = JA = kJ$.

Note 1. The definition of the incidence matrix of a block design varies with different authors and even with different works of the same author. Some say that the matrix designated by A^T above is the incidence matrix instead of A. Since symmetric block designs yield normal incidence matrices this results in no errors for the designs themselves. But it is sometimes a factor in the mechanics of the proofs. Thus one should be careful to check the incidence matrix definition being used when consulting a reference.

Note 2. If a, b, c are squarefree integers which are relatively prime in pairs and not all of the same sign, the diophantine equation $ax^2 + by^2 + cz^2 = 0$ has a solution in integers, not all zero, if and only if $-bc, -ac, -ab$ are quadratic residues of a, b, c respectively. [If $(g,m) = 1$, g is said to be a quadratic residue of m if there exists an integer x such that $x^2 \equiv g$ modulo m and called a quadratic nonresidue otherwise.] Given any diophantine equation $ax^2 + by^2 + cz^2 = 0$, there is an associated diophantine equation whose coefficients satisfy the restrictions above. Clearly any square factor may be dropped from a, b, c without significant change. Similarly any factor common to a, b, c can be divided out. Thus assume a, b, c are squarefree with greatest common divisor 1, and assume $(a,b) = g > 1$. Then, if the equation has a solution g must divide cz^2, but since a, b are squarefree and $(a, b, c) = 1$ this implies that g divides z. Thus the equation becomes $ax^2 + by^2 + cg^2z_1^2 = 0$. Thus g may be divided out. Clearly $(a,c) > 1$ or $(b,c) > 1$ can be handled in the same manner yielding an equation of the desired type.

C. Integral Solutions to the Incidence Equation

Since no further existence tests for difference sets or more general configurations are discussed in this section, the reader may wish to skip to section II.D. The material cited here not only is of independent interest but also represents a survey of the present status of the integral solution problem; one would hope that a complete solution of this problem would provide new existence criteria for block designs and difference sets.

As noted previously there is a natural correspondence between quadratic forms and symmetric matrices which links each form

$$f(x) = \sum_{i,j=1}^{n} c_{ij}x_i x_j \qquad c_{ij} = c_{ji} \tag{2.11}$$

with the matrix of its coefficients $C = (c_{ij})$. In this discussion the coefficients are restricted to the real numbers. Given two n-variable forms $f(x)$, $g(y)$ and their matrices C, D, $f(x)$ is said to <u>represent</u> $g(y)$ if there exists a linear transformation taking $f(x)$ into $g(y)$. That is, if there exists a substitution

$$x_i = \sum_{j=1}^{n} s_{ij}y_j \qquad i = 1,\ldots,n$$

transforming $f(x)$ into $g(y)$. If $S = (s_{ij})$, this means that

$$S^T CS = D \tag{2.12}$$

in which case C is said to <u>represent</u> D. If S is restricted to rational coefficients then $f(x)$ is said to represent $g(y)$ <u>rationally</u> (C represents D <u>rationally</u>) with analogous statements for other coefficient restrictions. Further, $f(x)$ represents $g(y)$ <u>rationally without essential denominator</u> if, for every integer m, there is a matrix S of rank n such that $S^T CS = D$, where S has rational elements with denominators prime to m. If two forms $f(x)$, $g(y)$ represent each other rationally without essential denominator they are called <u>semi-equivalent</u> and said to belong to the same <u>genus</u>.

If $f(x)$ represents $g(y)$ integrally and the transformation matrix S has determinant ± 1, then clearly $g(y)$ represents $f(x)$ integrally with transformation matrix of determinant ± 1. Two such forms are called <u>equivalent</u> and said to belong to the same <u>class</u>. Thus equivalent forms are clearly semi-equivalent; in fact a genus will in general contain several classes of forms.

Turning back to block designs again, consider the incidence equation

$$A^T A = (k - \lambda)I + \lambda J = B . \qquad (2.13)$$

This says no more than that the form $f(x) = x_1^2 + \cdots + x_v^2$ (associated with the $v \times v$ identity matrix I) represents the form $g(y) = (k - \lambda)(y_1^2 + \cdots + y_v^2) + \lambda(y_1 + \cdots + y_v)^2$ with a 0, 1 transformation matrix A.

The Bruck-Ryser-Chowla Theorem (Theorem 2.1 above) establishes necessary and sufficient conditions for the existence of a rational matrix A satisfying equation (2.13). Unfortunately the theory of integral representations of quadratic forms is not yet complete, even though a great deal of work has been done on such problems [see Jones (1950), Watson (1960) and O'Meara (1963) for this work.] However the main concern here is equation (2.13), not all possible integral representation problems, thus a complete theory is not required for these purposes.

Goldhaber (1960) studied the integral representation problem posed by equation (2.13) for a restricted set of parameters v, k, λ. He proved

Theorem 2.2. If v, k, λ satisfy the Bruck-Ryser-Chowla theorem and $(k,n) = 1$, then there exists a form in the genus of I which represents B integrally.

However, for $v > 8$, the genus of the identity matrix contains more than one equivalence class of forms (Magnus, 1937), thus no immediate conclusion can be drawn regarding integral representation from this result.

A 0, 1 matrix A satisfying equation (2.13) with $k(k - 1) = \lambda(v - 1)$ is the incidence matrix of a symmetric block design, although this is not immediately apparent. For, with the incidence matrix definition of section I.C. above (i.e., $a_{ij} = 1$ if and only if block i contains object j), equation (2.13) yields the fact that every object occurs in exactly k blocks and that any pair of objects occurs in exactly λ distinct blocks. So, the fact that every block contains precisely k objects is yet to be established. Let blocks $B_0, \ldots, B_{v-1}$ contain $b_0, \ldots, b_{v-1}$ objects respectively. Then, from the conclusions already drawn, it follows that

$$b_0 + b_1 + \cdots + b_{v-1} = kv$$

$$b_0(b_0 - 1) + b_1(b_1 - 1) + \cdots + b_{v-1}(b_{v-1} - 1) = \lambda v(v - 1) = vk(k - 1) .$$

So

$$\sum_{i=0}^{v-1} (b_i - k) = 0 = \sum_{i=0}^{v-1} [b_i(b_i - 1) - k(k - 1)]$$

from which it follows that

$$\sum_{i=0}^{v-1} (b_i - k)^2 = \sum_{i=0}^{v-1} [b_i(b_i - 1) - k(k - 1) + b_i + 2k^2 - k - 2kb_i] = 0 .$$

But $\Sigma(b_i - k) = 0 = \Sigma(b_i - k)^2$ implies that $b_i = k$ for all i. That is, every block contains precisely k objects, as was to be proved.

Note that if A satisfies equation (2.13) then so does $-A$. Of course, only one of these can be a 0, 1 matrix. Ryser (1952) has shown:

Theorem 2.3. If a normal integral matrix A satisfies equation (2.13), for $k(k - 1) = \lambda(v - 1)$, then A or $-A$ is a 0, 1 matrix and hence is the incidence matrix of a symmetric block design.

As pointed out in section I.C. above, the incidence matrix of a symmetric block design is normal; thus it satisfies:

(i) $A^T A = (k - \lambda)I + \lambda J$

(ii) $AA^T = (k - \lambda)I + \lambda J$

(iii) $AJ = kJ$

(iv) $JA = kJ$

Ryser has shown that a nonsingular $v \times v$ matrix, satisfying one of (i), (ii) and also one of (iii), (iv) satisfies all four of them and also the relation $k(k - 1) = \lambda(v - 1)$. [In Chapter V, the fact that (ii), (iii) imply (i), (iv) is needed; so a proof of this much of Ryser's result is included here. Since

$|B| = (k - \lambda)^{v-1}(v\lambda - \lambda + k)$, the nonsingularity of A means that $k - \lambda \neq 0$, $v\lambda - \lambda + k \neq 0$. By assumption

$$AA^T = (k - \lambda)I + \lambda J , \qquad AJ = kJ .$$

Multiplying this last by A^{-1} shows that $k \neq 0$ and that $A^{-1}J = k^{-1}J$. Whereas transposing both sides of this same equation yields $JA^T = kJ$, since $J = J^T$. Further, note that $J^2 = vJ$. So

$$A^T = A^{-1}(AA^T) = (k - \lambda)A^{-1} + \lambda A^{-1}J = (k - \lambda)A^{-1} + \lambda k^{-1}J$$

$$kJ = JA^T = (k - \lambda)JA^{-1} + \lambda k^{-1}J^2 = (k - \lambda)JA^{-1} + \lambda k^{-1}vJ .$$

Thus,

$$JA^{-1} = \frac{kJ - \lambda k^{-1}vJ}{k - \lambda} = \left[\frac{k - \lambda k^{-1}v}{k - \lambda}\right] J = mJ$$

which serves to define m. Thus $J = mJA$ and

$$vJ = J^2 = (mJA)J = mJ(AJ) = mJ(kJ) = mkvJ$$

which provides $mk = 1$ or $m = k^{-1}$. From $JA^{-1} = mJ$ and $m = k^{-1}$ it follows that $JA^{-1} = k^{-1}J$. Thus $J = k^{-1}JA$ or $JA = kJ$; that is (iv) has been established. As shown above $A^T = (k - \lambda)A^{-1} + \lambda k^{-1}J$; multiplying through by A yields

$$A^TA = (k - \lambda)I + \lambda k^{-1}JA = (k - \lambda)I + \lambda J$$

by (iv). But this is (i). So (ii), (iii) have been shown to imply (i), (iv) as promised.] Using a different part of Ryser's result, note that an <u>integral</u> solution of equation (2.13) which also satisfies one of $AJ = kJ$, $JA = kJ$ is normal and by Theorem 2.3 must be the incidence matrix of a symmetric block design.

Let A be a matrix satisfying $XX^T = B$, then if any column of A is multiplied through by -1, the resultant matrix C still satisfies $XX^T = B$. Thus, given such a matrix A, there is no loss of generality in assuming that its column sums are nonnegative.

Theorem 2.4. (Ryser, 1952) If A is an integral matrix, with nonnegative column sums, satisfying $AA^T = B$ with $k(k - 1) = \lambda(v - 1)$, if (k, λ) is square-free and if $k - \lambda$ is odd, then A is the incidence matrix of a symmetric block design.

The condition $k - \lambda$ odd is necessary, for many counter examples exist when $n = k - \lambda$ is even and $\lambda = 1$. In particular [see Hall (1967), p. 286].

Theorem 2.5. Let $v = n^2 + n + 1$, $k = n + 1$, $\lambda = 1$ and let A be an integral $v \times v$ matrix with nonnegative column sums satisfying $AA^T = B$, then A is of one of two types: (i) A is the incidence matrix of a block design or (ii) n is even, one of the columns of A has a single zero and $n^2 + n$ entries $+1$, $n + 1$ columns consist of $n + 1$ entries $+1$ and n^2 zeros, and the remaining columns have sum zero. There exists an A of type (ii) for every n which is the order of an Hadamard matrix and also for $n = 10$.

The orders of Hadamard matrices (1, -1 square matrices of order n with determinants $\pm n^{n/2}$) are limited to $n = 1, 2$ and $4t$ for positive integers t. It is not known whether they exist for all t [see the note at the end of section IV. B. for more information on this subject].

In addition to the solutions of type (ii) given above, E. C. Johnson (1966A) has found infinitely many type (ii) solutions for the case $n \equiv 2 \pmod 4$. Further, when $n = 2^a$ there are known to be both types of solutions; for such block designs always exist as well as Hadamard matrices of order $n = 2^a$. It has been conjectured that type (ii) solutions always exist for even n whenever $v = n^2 + n + 1$, $k = n + 1$, $\lambda = 1$ satisfy the Bruck-Ryser-Chowla condition.

D. The Theorem of Hall and Ryser

If a cyclic difference set exists, then its Hall-polynomial $\theta(x) = x^{d_1} + \cdots + x^{d_k}$ satisfies the congruence

$$\theta(x)\theta(x^{-1}) \equiv n + \lambda(1 + x + \cdots + x^{v-1}) \qquad (\text{mod } x^v - 1) \tag{2.14}$$

Thus, for any divisor w of v, one has

$$\theta(x) \equiv b_0 + b_1 x + \cdots + b_{w-1}x^{w-1} \qquad (\text{mod } x^w - 1) \tag{2.15}$$

$$\theta(x)\theta(x^{-1}) \equiv n + \frac{\lambda v}{w}(1 + x + \cdots + x^{w-1}) \qquad (\text{mod } x^w - 1) \tag{2.16}$$

where b_i is the number of d_j in D satisfying $d_j \equiv i \text{ mod } w$. This implies [equate coefficients in congruence 2.16]

$$\sum_{i=0}^{w-1} b_i^2 = n + \frac{\lambda v}{w}, \qquad \sum_{i=0}^{w-1} b_i b_{i-j} = \frac{\lambda v}{w}$$

for $j = 1,\ldots,w - 1$. (Here the subscript $i - j$ is taken modulo w.) So if S is the cyclic matrix defined by

$$S = \begin{bmatrix} b_0 & b_1 & \cdots & b_{w-1} \\ b_1 & b_2 & \cdots & b_0 \\ \cdot & \cdot & & \cdot \\ \cdot & \cdot & & \cdot \\ \cdot & \cdot & & \cdot \\ b_{w-1} & b_0 & \cdots & b_{w-2} \end{bmatrix}$$

it follows that

$$SS^T = D \tag{2.17}$$

where D has $n + (\lambda v/w)$ on the main diagonal and $\lambda v/w$ in all other positions. Applying the method of Chowla and Ryser (see the proof of Theorem 2.1 above), it follows that in order for there to exist a rational matrix S of odd order w satisfying equation (2.17) it is necessary that the equation

$$z^2 = nx^2 + (-1)^{(w-1)/2}(\lambda v/w)\, y^2 \tag{2.18}$$

have a solution in integers x, y, z, not all zero. If n is not a square, then in this solution $y \neq 0$. Thus, for this case, the integers $\bar{z} = zk + nx$, $\bar{x} = z + kx$ and $\bar{y} = y(\lambda v/w)$ are not all zero and they satisfy the diophantine equation

$$\bar{z}^2 = n\bar{x}^2 + (-1)^{(w-1)/2} w\bar{y}^2 . \tag{2.19}$$

Of course this equation always has a non-trivial solution when n is a square. Thus

Theorem 2.6 (Hall and Ryser, 1951) If a non-trivial difference set exists for odd v, then, for every divisor w of v, equation (2.19) has a solution in integers $\bar{x}, \bar{y}, \bar{z}$, not all zero.

From the derivation above it is obvious that equation (2.19) has a non-trivial solution for $w = v$ whenever condition (ii) of the Bruck-Ryser-Chowla Theorem is satisfied. Thus Theorem 2.6 can be stronger than the Bruck-Ryser-Chowla Theorem only when v is odd and composite. An example is furnished by the parameter set $v, k, \lambda = 39, 19, 9$. Here with $w = v = 39$, the two equations are $z^2 = 10x^2 - 9y^2$ and $z^2 = 10x^2 - 39y^2$ and they have solutions $x, y, z = 1, 1, 1$ and $2, 1, 1$ respectively. However, for $w = 13$, the equation of Hall and Ryser is $z^2 = 10x^2 + 13y^2$ and it has no non-trivial integral solutions. Thus no cyclic difference set exists with parameters 39, 19, 9. [A symmetric block design with these parameters does exist however. It is easily derived from any Hadamard matrix of order 40; Hall (1967, Chapter 14) discusses construction methods for

Hadamard matrices.]

Hughes (1957) used the work of Hall and Ryser to prove a more general theorem about the structure of certain symmetric block designs. However, for difference sets the result of Hughes says no more than Theorem 2.6 above.

Note. Again only the conditions relating to rational solution of equation (2.10) were developed, whereas it was known that S was integral, cyclic and that $b_i \leq v/w$ for $i = 0,1,\ldots,w-1$. In particular perhaps some improvement in the status of the integral representation problem for quadratic forms (see section II.C) would provide stricter existence tests.

E. Results of Mann, Rankin, Turyn and Yamamoto

By far the most stringent difference set existence tests yet discovered arise from algebraic number theoretic consideration of the equation

$$\theta(\zeta_v)\,\theta(\zeta_v^{-1}) = n \tag{2.20}$$

which must hold for any v^{th} root of unity $\zeta_v \neq 1$ if a difference set is to exist. Most of the results cited below are attributed to Mann (1964), Turyn (1965) and Yamamoto (1963) even though the work in Mann (1952, 1965, 1967), Rankin (1964) and Turyn (1960, 1961) is directly related. In fact many of the results were discovered independently by more than one of these authors. A few of the theorems require the knowledge of a multiplier or w-multiplier for the hypothetical difference set under question (see sections I.E. and I.F. for these definitions). Chapter III discusses conditions on the parameters under which certain multipliers are known to exist.

The results of this section are all stated in terms of the parameters v, k, λ, n and their divisors; thus no algebraic number theory is required to either understand the nature of the results or to apply them. However, some knowledge of algebraic number theory is required for the proofs. In order to minimize this requirement some basic facts about cyclotomic fields (i.e., algebraic number fields generated by the roots of unity) are presented at the end of this section (Theorems 2.18, 2.19 and 2.20). They will be referred to when they are used. Unless otherwise specified, ζ_d will be an arbitrary primitive d^{th} root of unity and $K(\zeta_d)$ will denote the cyclotomic field it generates over the rational field.

If a, b, c are integers $(c \geq 0)$ and a^c divides b while a^{c+1} does not, then a^c is said to <u>strictly divide</u> b. The same terminology is used if a and b are ideals.

<u>Theorem 2.7</u> (Mann, 1964) Let $w > 1$ be a divisor of v and assume a non-trivial v, k, λ-difference set exists with w-multiplier $t \geq 1$. Let p be a prime divisor of n for which $(p,w) = 1$. If there exists an integer $f \geq 0$ such that $tp^f \equiv -1$ modulo w, then n is strictly divisible by an even power of p.

<u>Proof</u>. Let P be a prime ideal divisor of p in $K(\zeta_w)$ and let P^i strictly divide $\theta(\zeta_w)$. Then by (2.47) of Theorem 2.19 below and equations (1.14), (2.20) it follows that P^i also strictly divides

$$\theta(\zeta_w^t), \qquad \theta(\zeta_w^{tp}), \qquad \theta(\zeta_w^{-1}).$$

Thus, by equation (2.20), P^{2i} strictly divides n and this implies (Theorem 2.19) that n is strictly divisible by an even power of p.

This theorem, which (as shown above) is an almost trivial consequence of the prime ideal structure of $\theta(\zeta_w)$ is an extremely important result. For, as far as cyclic difference sets are concerned, it contains both the Bruck-Ryser-Chowla theorem (odd v) and the theorem of Hall and Ryser. In order to establish the dependence of these results on Theorem 2.7 it is convenient to know the connection between diophantine equations of the type

$$x^2 = \alpha y^2 + \beta z^2 \qquad \alpha, \beta \text{ integers} \tag{2.21}$$

and the so-called <u>Hilbert norm residue symbol</u> $(\alpha,\beta)_p$. The basic fact is that equation (2.21) has a solution in integers x, y, z, not all zero, if and only if, $(\alpha,\beta)_p = +1$ for all finite primes p and α, β are not both negative [this last fact is written $(\alpha,\beta)_\infty = +1$]. See Jones (1950, p. 26 ff.) for this as well as the other properties of the Hilbert symbol which are needed. These are ($p = \infty$ allowed):

$$(\alpha,\beta)_p = \pm 1 \tag{2.22}$$

$$(\alpha,\beta)_p = (\beta,\alpha)_p \tag{2.23}$$

$$(\alpha,\beta)_p(\alpha,\gamma)_p = (\alpha,\beta\gamma)_p \tag{2.24}$$

Corollary 2.8 (Yamamoto, 1963) Assume that a v, k, λ - difference set exists and let $q^* = (-1)^{(q-1)/2}\, q$ for an odd divisor q of v. Then if r is a prime such that r^e strictly divides n, it follows that the equation

$$r^e\, x^2 + (-1)^{(q-1)/2}\, qy^2 = z^2 \tag{2.25}$$

has a solution in integers x, y, z, not all zero. Or, equivalently, that the Hilbert norm residue symbol

$$(r^e, q^*)_p = +1 \tag{2.26}$$

for all primes p.

Proof. Since $(q_1 q_2)^* = q_1^* q_2^*$ it follows from (2.24) that q may be assumed to be a prime. If e is even, equation (2.25) has a solution $x,y,z = 1,0,r^{e/2}$. Thus attention may be restricted to odd $e = 2j + 1$. In this case (2.25) has a solution $(x,y,z = a, br^j, cr^j)$ whenever the equation

$$rx^2 + (-1)^{(q-1)/2} q\, y^2 = z^2 \tag{2.27}$$

has a solution $x,y,z = a,b,c$. Now $x,y,z = 1,1,0$ is a solution of (2.27) when $r = q \equiv 3(4)$. If $r = q \equiv 1(4)$ then since such a q has a representation (see

Nagell, 1951) $q = s^2 + t^2$ where s and t are integers, it follows that $x,y,z = s,t,q$ is a solution of (2.27). For $r \neq q$ Legendre's test (section II.B. note 2) requires r, q^*, $-rq^*$ to be quadratic residues of q, r, 1 respectively. Of these conditions, the last is trivial and the second follows from the first by quadratic reciprocity (Nagell, 1951). As for the first, if r were not a quadratic residue of q then, by Euler's criterion, $r^{(q-1)/2} \equiv -1 \pmod q$ and hence (Theorem 2.7) e would be even. This contradiction establishes the existence of a non-trivial solution for equation (2.27) and hence for equation (2.25), i.e., the corollary has been established.

In terms of the Hilbert symbol, the theorem of Hall and Ryser becomes $(n,w^*)_p = +1$ for all odd divisors w of v and this is an immediate consequence of equations (2.23), (2.24), (2.26). Using this for $w = v$, v odd, yields

$$+1 = (n,v^*)_p = (n,v)_p \, (n,(-1)^{(v-1)/2})_p$$

while $k^2 = \lambda v + n$ provides

$$+1 = (n,\lambda v)_p = (n,\lambda)_p (n,v)_p \, .$$

Together these equations show that

$$(n,\lambda)_p = (n,v)_p = (n,(-1)^{(v-1)/2})_p \, .$$

Hence

$$(n,\lambda)_p \, (n,(-1)^{(v-1)/2})_p = (n,(-1)^{(v-1)/2}\lambda)_p = +1 \, .$$

I.e., the Bruck-Ryser-Chowla condition for odd v has been established as a direct consequence of Theorem 2.7.

In particular, if $v,k,\lambda = 241,\ 16,\ 1$ the Bruck-Ryser-Chowla equation

$z^2 = 15x^2 + y^2$ has a solution $z = y = 1$, $x = 0$. As 241 is prime and the Hall-Ryser condition is only stronger for composite v, it must also be satisfied. However $3^{60} \equiv -1$ modulo 241; thus Theorem 2.7 shows that no difference set exists.

Another result of Mann (1964) is:

Theorem 2.9. Let t be a w-multiplier of a non-trivial v, k, λ - difference set for some divisor $w > 1$ of v. Let $t^f \equiv -1$ modulo w for some integer f. Then n is a square or $n = n_0^2 q^3$ for some prime q. In this later case

(i) $q \equiv 1$ modulo 4

(ii) v is odd

(iii) $w = q^s$, $s \geq 1$

(iv) t is a quadratic residue of q

(v) if $q_1 \neq q$ is another prime divisor of v then $t_1^{f_1} \not\equiv -1$ modulo q_1 for all multipliers t_1 and all integers f_1.

Proof. Since t is w-multiplier, so is t^f; thus $\theta(\zeta_w^{-1}) = \zeta_w^r\, \theta(\zeta_w)$ for some integer r. Then

$$[\theta(\zeta_w)]^2 = n\, \zeta_w^{-r} \; . \tag{2.28}$$

From this and the prime ideal factorization of n in the field of wth roots of unity (Theorem 2.19), it follows that $n = n_1^2$ or $n = n_1^2 n_2$ where every prime divisor of n_2 also divides w and n_2 may be assumed to be square free.

Unless $n = n_1^2 n_2$ the theorem is proved, so from now on assume $n = n_1^2 n_2$ with $n_2 > 1$. By Theorem 2.1 it follows that v and hence w and n_2 are necessarily odd. Let q be a prime divisor of n_2 and apply the same process as above with the odd prime q playing the role of w; this yields $n = m^2 q$. Thus $m^2 = n_1^2$ and $n_2 = q$ necessarily. If w were dq^s with $d > 1$ and $(d, q^s) = 1$, then the same process with d for w would show $q = 1$ which contradicts the assumption $n_2 > 1$. So $w = q^s$ $(s \geq 1)$ and $n = n_1^2 q$ with q prime. As $k^2 = \lambda v + n$, $n = k - \lambda$, it follows from $n \equiv v \equiv 0$ modulo q that q^2

divides n; thus with $n_0 = n_1/q$ the desired form $n = n_0^2 q^3$ emerges, where it is not assumed that $(n_0, q) = 1$.

As before, in $K(\zeta_q)$ there is an equation analogous to (2.28), namely

$$[\theta(\zeta_q)]^2 = n\,\zeta_q^{\ell} \tag{2.29}$$

for some integer $\ell \geq 0$. By definition $q^* = (-1)^{(q-1)/2} q$ and it was shown by Gauss (see Nagell, 1951, section 53) that with $\zeta_q = e^{2\pi i/q}$

$$\sqrt{q^*} = 1 + \zeta_q + \zeta_q^4 + \zeta_q^9 + \cdots + \zeta_q^{(q-1)^2}. \tag{2.30}$$

Hence $\sqrt{q^*}$ is an algebraic integer of $K(\zeta_q)$. [It is invariant under the field automorphism defined by $\zeta_q \to \zeta_q^t$, whenever t is a non-zero square modulo q. Further, if t is not a square modulo q, then the mapping $\zeta_q \to \zeta_q^t$ takes $\sqrt{q^*}$ into $-\sqrt{q^*}$.] Since $\sqrt{q^*}$ is an algebraic integer of $K(\zeta_q)$ it follows that the fraction $\Psi = \theta(\zeta_q)/n_1\sqrt{q^*}$ is an element of this field. Indeed, since

$$\left[\frac{\theta(\zeta_q)}{n_1\sqrt{q^*}}\right]^{4q} = \left[\frac{n\zeta_q^{\ell}}{n(-1)^{(q-1)/2}}\right]^{2q} = \left[(-1)^{(q-1)/2}\right]^2 = 1 \tag{2.31}$$

i.e., since Ψ satisfies the equation $x^{4q} = 1$, it follows that it is not only an algebraic integer but also a root of unity in $K(\zeta_q)$. But all roots of unity in $K(\zeta_q)$ satisfy $x^{2q} = 1$; thus it follows from equation (2.31) that $q \equiv 1$ modulo 4. As Ψ is a $2q^{th}$ root of unity, it is $\pm\eta$ for some q^{th} root of unity η. Thus

$$\theta(\zeta_q) = \pm\,\eta\, n_1\sqrt{q^*} \tag{2.32}$$

and as t is a q-multiplier it follows from congruence 1.14 that $\theta(\zeta_q^t) = \zeta_q^s\,\theta(\zeta_q)$. Now if t is not a square modulo q, it follows from this equation together with

(2.32) that

$$\mp \eta^t n_1 \sqrt{q^*} = \zeta_q^s(\pm \eta n_1 \sqrt{q^*}) \tag{2.33}$$

since, as noted above, $\zeta_q \to \zeta_q^t$ maps $\sqrt{q^*}$ into $-\sqrt{q^*}$. But (2.33) is a contradiction, so t necessarily is a non-zero square modulo q.

Of the conclusions of Theorem 2.9 only the last one remains to be verified. If the prime $q_1 \neq q$ were another (necessarily odd) prime divisor of v, for which there existed a multiplier t_1 and integer f_1 such that $t_1^{f_1} \equiv -1$ modulo q_1, then, by the process used above, it would follow that n was a square (since it cannot be both $n_0^2 q^3$ and $m_0^2 q_1^3$). This contradiction completes the proof of Theorem 2.9.

Note that neither $t_1 = t$ nor q dividing n_0 is excluded by this theorem. Consider the parameters $v, k, \lambda = 813, 29, 1$. Theorem 1.1 (section I.E) shows that 2 is a multiplier, hence also a q-multiplier for $q = 3$. Thus Theorem 2.9 shows that no such difference set exists.

Before establishing any further existence tests it is convenient to have some results concerning congruence relations in cyclotomic fields. Let C be a number theoretic function (i.e., a function which is zero except on the integers) and define the difference operator $\Delta(\rho)$ by

$$\Delta(\rho)\, C(i) = C(i + \rho) - C(i)$$

where ρ is a rational number not necessarily an integer. If an integer n exists such that $\Delta(n)\, C(i) = 0$ for all i, then call C periodic of period n.

<u>Theorem 2.10</u> (Yamamoto, 1963) Let $N = p_1^{\ell_1} \cdots p_s^{\ell_s}$ be the prime power decomposition of N. Let m be relatively prime to N, let C be a periodic number theoretic function with period N whose values $C(i)$ are algebraic integers of the cyclotomic field $K(\zeta_m)$ and let $f(x) = \Sigma_{i=0}^{N-1} C(i)x^i$. Further, let d be a divisor of N and let α be an integer of $K(\zeta_m)$.

Then, in order that $f(\zeta_N^r) \equiv 0 \pmod{\alpha}$ for all divisors r of d, it is necessary and sufficient that

$$p_1^{t_1} \cdots p_s^{t_s} \Delta(Np_1^{-t_1-1}) \cdots \Delta(Np_s^{-t_s-1})\, C(i) \equiv 0 \pmod{\alpha}$$

for all i and for all $t_1,\ldots,t_s$ such that $p_1^{t_1} \cdots p_s^{t_s}$ divides d.

Proof. (1) Let $s = 1$, $N = p^\ell$, $d = p^u$ and, proceeding by induction on u, let $u = 0$. Now

$$f(\zeta_N) = \sum_{i=0}^{p^{\ell-1}-1} \sum_{j=0}^{p-1} C(i + jp^{\ell-1})\, \zeta_N^{i+jp^{\ell-1}}$$

$$= \sum_{i=0}^{p^{\ell-1}-1} \sum_{j=1}^{p-1} [C(i + jp^{\ell-1}) - C(i)]\, \zeta_N^{i+jp^{\ell-1}}$$

since $\zeta_N^{p^{\ell-1}} = \zeta_p$ is a primitive p^{th} root of unity. The $\varphi(p^\ell)$ integers $\zeta_N^{i+jp^{\ell-1}}$ for $0 \leq i < p^{\ell-1}$, $1 \leq j < p$ form an integral basis for $K(\zeta_{mN})$ over $K(\zeta_m)$; thus $f(\zeta_N) \equiv 0 \pmod{\alpha}$ if and only if $C(i + jp^{\ell-1}) - C(i) \equiv 0 \pmod{\alpha}$ for all these values of i and j. This condition is equivalent to the desired one

$$\Delta(Np^{-1})\, C(i) \equiv 0 \pmod{\alpha}$$

for all i. Thus the theorem is proved for $s = 1$ and $u = 0$.

(2) Let $s = 1$, $u > 0$ and assume that the theorem is true for all integers t, $0 \leq t < u$. Let

$$g(x) = p \sum_{i=0}^{p^{\ell-1}-1} C(i)\, x^{ip}$$

and note that

$$f(x^p) \equiv \sum_{i=0}^{p^{\ell}-1} C(i)x^{ip} \equiv \sum_{i=0}^{p^{\ell-1}-1} \left[\sum_{j=0}^{p-1} C(i + jp^{\ell-1}) \right] x^{ip} \quad (\text{mod } x^N - 1) .$$

Since $f(\zeta_N) \equiv 0 \pmod{\alpha}$ implies [see part (1) above] that $C(i) \equiv C(i + jp^{\ell-1})$ modulo α it follows that

$$f(x^p) \equiv g(x) \quad \text{mod}(\alpha, x^N - 1) .$$

Thus the condition $f(\zeta_N^{p^t}) \equiv 0 \pmod{\alpha}$ for all t such that $0 \le t \le u$ is equivalent to $f(\zeta_N^{p^t}) \equiv 0 \pmod{\alpha}$ and $g(\zeta_N^{p^t}) \equiv 0 \pmod{\alpha}$ for all t such that $0 \le t \le u - 1$. By the induction hypothesis these last two congruences are equivalent to $p^t \Delta(p^{\ell-t-1})\, C(i) \equiv 0 \pmod{\alpha}$ and $p^{t+1} \Delta(p^{\ell-t-2})\, C(i) \equiv 0 \pmod{\alpha}$ for all t such that $0 \le t \le u - 1$, that is, to $p^t \Delta(p^{\ell-t-1})\, C(i) \equiv 0 \pmod{\alpha}$ for all t such that $0 \le t \le u$. Thus, the theorem has been proved for $s = 1$.

(3) Now assume $s > 1$ and assume the validity of the theorem for smaller values of s. Put $N = N_1 N'$, $N_1 = p_1^{\ell_1}$, $N' = p_2^{\ell_2} \cdots p_s^{\ell_s}$, $d = d_1 d'$, $d_1 = (N_1, d)$, $d' = (N', d)$. Any divisor r of d can be uniquely written as $r = r_1 r'$ where r_1 divides d_1 and r' divides d'. For any integer i there exist integers j, k such that $i \equiv N'j + N_1 k \pmod{N}$ with j determined modulo N_1 and k determined modulo N'. Hence

$$f(x) \equiv \sum_{j=0}^{N_1-1} \sum_{k=0}^{N'-1} C(N'j + N_1 k)\, x^{N'j+N_1 k} \quad (\text{mod } x^N - 1)$$

$$f(\zeta_N^r) = \sum_{j=0}^{N_1-1} \sum_{k=0}^{N'-1} C(N'j + N_1 k)\, \zeta_N^{N'jr}\, \zeta_N^{N_1 kr} = \sum_{j=0}^{N_1-1} C^*(\zeta_N^{N_1 r}, j)\, \zeta_N^{N'rj}$$

$$\text{where} \quad C^*(y,j) = \sum_{k=0}^{N'-1} C(N'j + N_1k)y^k .$$

Now $\zeta_N^{N'} = \xi$ is a primitive N_1^{th} root of unity and $\zeta_N^{N_1} = \eta$ is a primitive N'^{th} root of unity. Further, the condition $f(\zeta_N^r) \equiv 0 \pmod{\alpha}$ for some ζ_N implies the same congruence for all primitive N^{th} roots of unity. Thus $f(\zeta_N^r) \equiv 0 \pmod{\alpha}$ for some ζ_N implies that

$$\sum_{j=0}^{N_1-1} C^*(\eta^{r'},j)\, \xi^{r_1 j} \equiv 0 \qquad \pmod{\alpha}$$

for all primitive N_1^{th} roots of unity ξ and for all primitive N' roots of unity η. Note that the $C^*(\eta^{r'},j)$ are algebraic integers of $K(\zeta_{mN'})$ and that $(mN',N_1) = 1$. Thus applying the theorem for $s = 1$ to this case, i.e., to the polynomial

$$\sum_{j=0}^{N_1-1} C^*(\eta^{r'},j)\, x^j$$

yields the result that $f(\zeta_N^r) \equiv 0 \pmod{\alpha}$ for all divisors r of d if and only if

$$p_1^{t_1} \Delta_j(N_1 p_1^{-t_1-1})\, C^*(\eta^{r'},j) \equiv 0 \qquad \pmod{\alpha} \qquad (2.34)$$

for all t_1 such that $p_1^{t_1}$ divides d_1 and for all divisors r' of d'. Here $\Delta_j(\rho)$ indicates that the difference operator applies to the argument j.

Congruence 2.34 may be rewritten as

$$p_1^{t_1} \sum_{k=0}^{N'-1} \Delta_j(N_1 p_1^{-t_1-1})\, C(N'j + N_1 k)\eta^{r'k} \equiv 0 \qquad (\text{mod } \alpha) \qquad (2.35)$$

for all t_1 and r' such that $p_1^{t_1}$ divides d_1 and r' divides d'. Apply the induction hypothesis to the polynomial

$$p_1^{t_1} \sum_{k=0}^{N'-1} \Delta_j(N_1 p_1^{-t_1-1})\, C(N'j + N_1 k)\, x^k$$

which can be done since N' has $s - 1$ distinct prime divisors, since the coefficients of this polynomial are algebraic integers of $K(\zeta_m)$ and since $(m,N') = 1$. Thus congruence 2.35 holds if and only if

$$p_1^{t_1} p_2^{t_2} \cdots p_s^{t_s} \Delta_j(N_1 p_1^{-t_1-1})\, \Delta_k(N'p_2^{-t_2-1}) \cdots \Delta_k(N'p_s^{-t_s-1})\, C(N'j + N_1 k)$$

$$= p_1^{t_1} \cdots p_s^{t_s} \Delta(Np_1^{-t_1-1}) \cdots \Delta(Np_s^{-t_s-1})\, C(i) \equiv 0 \qquad (\text{mod } \alpha)$$

for all i and for all $t_1,\ldots,t_s$ such that $p_1^{t_1} \cdots p_s^{t_s}$ divides d. That is, the theorem has been established.

Folliwng Mann (1967) note that:

Corollary 2.11. Let $\alpha > 0$ be an integer and let C be an integer valued, periodic, number theoretic function of period N, whose values $C(i)$ satisfy $0 \leq C(i) \leq M$ and

$$0 \not\equiv \sum_{i=0}^{N-1} C(i)\zeta_N^i \equiv 0 \qquad (\text{mod } \alpha) .$$

Then, if N is the product of exactly s distinct prime powers, it follows that

$\alpha \leq 2^{s-1} M$.

Proof. Apply Theorem 2.10 with $m = d = 1$; this yields

$$\Delta(Np_1^{-1}) \ldots \Delta(Np_s^{-1})\ C(i) \equiv 0 \qquad (\text{mod } \alpha) \qquad (2.36)$$

which must hold for all integers i. For some i, say i_0, the left side of (2.36) must be non-zero. [For otherwise Theorem 2.10 could be applied to this congruence for any arbitrary value of α and this would contradict the condition $\Sigma\, C(i)\zeta_N^i \neq 0$.] Consider (2.36) at i_0. The left side contains 2^s terms $C(j)$ with $0 \leq C(j) \leq M$; exactly half of which have negative signs. Thus the magnitude of the left side of (2.36) is bounded by $2^{s-1}M$, i.e., $\alpha \leq 2^{s-1}M$ as was to be shown.

Corollary 2.11 will be put to good use in the proof of the next existence test. But first a lemma and a definition are needed.

Lemma 2.12. Let $A(x) = \sum_{i=0}^{w-1} a_i x^i$, where the a_i are algebraic integers of $K(\zeta_s)$, and let $(s,w) = (m,w) = 1$ for some integer m. If

$$A(\zeta_w^j) \equiv 0 \qquad \text{modulo } m \qquad (2.37)$$

for $0 \leq j \leq w - 1$, then $a_i \equiv 0$ modulo m for all i. If the a_i are rational integers and (2.37) holds only for $1 \leq j \leq w - 1$, then

$$a_0 \equiv a_1 \equiv \cdots \equiv a_{w-1} \qquad \text{modulo } m .$$

Proof. Assume that (2.37) holds for $0 \leq j \leq w - 1$, i.e., that

$$\begin{bmatrix} 1 & 1 & 1 & \cdots & 1 \\ 1 & \zeta_w & \zeta_w^2 & \cdots & \zeta_w^{w-1} \\ 1 & \zeta_w^2 & \zeta_w^4 & \cdots & \zeta_w^{w-2} \\ \cdot & \cdot & \cdot & \cdots & \cdot \\ \cdot & \cdot & \cdot & \cdots & \cdot \\ \cdot & \cdot & \cdot & \cdots & \cdot \\ 1 & \zeta_w^{w-1} & \zeta_w^{w-2} & \cdots & \zeta_w \end{bmatrix} \begin{bmatrix} a_0 \\ a_1 \\ a_2 \\ \cdot \\ \cdot \\ \cdot \\ a_{w-1} \end{bmatrix} = \begin{bmatrix} b_0 \\ b_1 \\ b_2 \\ \cdot \\ \cdot \\ \cdot \\ b_{w-1} \end{bmatrix} \tag{2.38}$$

where $b_i \equiv 0$ modulo m for all i. The determinant of the system of equations (2.38) is van der Monde and not zero, for it is known to be

$$\prod_{i>j} (\zeta_w^{i-1} - \zeta_w^{j-1}) = \prod_{i>j} \zeta_w^{j-1} (\zeta_w^{i-j} - 1) .$$

Since $F(x) = x^{w-1} + \cdots + x + 1 = (x - \zeta_w)(x - \zeta_w^2) \ldots (x - \zeta_w^{w-1})$ it follows (for $x = 1$) that

$$w = (1 - \zeta_w)(1 - \zeta_w^2) \ldots (1 - \zeta_w^{w-1}) .$$

This shows that only primes dividing w may occur in the determinant above. So, by Cramer's rule, it follows that $a_i \equiv 0$ modulo m for all i.

Suppose the a_i are rational integers and suppose that $A(1) = \sigma \not\equiv 0$ modulo m. Consider the polynomial $B(x) = A(x) + \ell F(x)$, where ℓ is chosen so that $\sigma + \ell w \equiv 0$ modulo m [certainly possible since $(m,w) = 1$]. Then $B(\zeta_w^j) \equiv 0$ modulo m for $0 \leq j \leq w - 1$ and hence (by the first part of this lemma) $a_i + \ell \equiv 0$ modulo m for all i, that is

$$a_0 \equiv a_1 \equiv \cdots \equiv a_{w-1} \qquad \text{modulo } m$$

as was to be proved.

Let p be a prime and let p^{ℓ} strictly divide the integer w, i.e., let $w = p^{\ell} w_1$ with $(p, w_1) = 1$. If there exists an integer $f > 0$ such that $p^f \equiv -1$ modulo w_1, then p is said to be <u>self-conjugate modulo w</u>. If all the prime divisors of an integer m are self-conjugate modulo w, then m is said to be <u>self-conjugate modulo w</u>. [From the prime ideal factorization of m in the wth roots of unity (see Theorem 2.19) it follows that, if m is self-conjugate modulo w then every prime ideal divisor of m in this field is fixed under complex conjugation. Thus, if a difference set is under consideration for which m^2 divides n, w divides v and m is self-conjugate modulo w, it would follow from $\theta(\zeta_w)\,\theta(\zeta_w^{-1}) = n$ that $\theta(\zeta_w) \equiv 0$ modulo m.] Note that if m is self-conjugate modulo w it is also self-conjugate modulo any divisor of w.

<u>Theorem 2.13</u> (Turyn, 1965) Assume a non-trivial v,k,λ-difference set exists. Let m^2 divide n and suppose that $m > 1$ is self-conjugate modulo w for some divisor $w > 1$ of v. If $(m,w) = 1$ then $m \leq (v/w)$. If $(m,w) > 1$ then

$$m \leq 2^{r-1}\,(v/w) \tag{2.38}$$

where r is the number of distinct prime factors of (m,w).

<u>Proof</u>. Since m is self-conjugate modulo w it follows (as noted above) that $\theta(\zeta_w^j) \equiv 0$ modulo m for $1 \leq j \leq w-1$.

Let $w = v_1 v_2$ where $(v_2, m) = 1$ and where every prime divisor of v_1 also divides m. Further, let

$$\theta(x) \equiv \sum_{i=0}^{w-1} a_i\, x^i \qquad \mod (x^w - 1)$$

where $0 \leq a_i \leq v/w$ necessarily. First consider the case $(m,w) = 1$, i.e., $w = v_2$. Here $\theta(\zeta_w^j) \equiv 0$ modulo m for $1 \leq j \leq w-1$ and by Lemma 2.12 it follows that all the coefficients a_i are congruent modulo m. Let α be the smallest such coefficient. Then

$$\sum_{i=0}^{w-1} a_i x^i = \alpha \sum_{i=0}^{w-1} x^i + \sum_{i=0}^{w-1} (a_i - \alpha)\, x^i$$

where $0 \leq a_i - \alpha \leq v/w$ and not all the $a_i - \alpha$ are zero. [For, if they were all zero, then $\theta(\zeta_w^j)$ and hence n would also be zero; a contradiction]. Since m divides $a_i - \alpha$ for all i it follows that $m \leq a_i - \alpha \leq v/w$ as was to be proved.

Assume now that $(m,w) > 1$ has exactly r distinct prime divisors. Consider the polynomial

$$\psi(x) = \sum_{i=0}^{w-1} a_i \zeta_{v_1}^i x^i = \sum_{j=0}^{v_2-1} b_j x^j$$

where, as usual, ζ_{v_1} is a primitive v_1^{st} root of unity and

$$b_j = a_j \zeta_{v_1}^j + a_{j+v_2} \zeta_{v_1}^{j+v_2} + \cdots + a_{j+(v_1-1)v_2} \zeta_{v_1}^{j+(v_1-1)v_2} .$$

Since m is self-conjugate modulo w and $v_1 > 1$ it follows that

$$\theta(\zeta_{v_1} \zeta_{v_2}^{\ell}) = \psi(\zeta_{v_2}^{\ell}) \equiv 0 \qquad (\text{mod } m)$$

for $0 \leq \ell \leq v_2 - 1$. By Lemma 2.12 this implies that $b_j \equiv 0$ modulo m for all j; further $b_j \neq 0$ for some j since $n \neq 0$. Applying Corollary 2.11 to such a b_j, yields

$$m \leq 2^{r-1}\, (v/w)$$

as was to be proved.

This theorem shows, for example, that if v is a prime power then $(n,v) = 1$. [For if p divides (n,v) then p^2 divides n, since $k^2 = \lambda v + n$ and $n = k - \lambda$. Thus the theorem may be applied with $m = p$, $w = v$.] In particular no cyclic difference set may have parameters $v,k,\lambda = 16, 6, 2$. Consider the parameter set $v,k,\lambda = 56, 11, 2$. Since $n = 9$ is a square, all previous theorems allow the possibility that a difference set might exist with these parameters. But it is ruled out by Theorem 2.13, since $3^3 \equiv -1$ modulo 28.

If $(a,c) = 1$, $a^b \equiv 1$ modulo c and $a^x \not\equiv 1$ modulo c for any x $(1 \leq x < b)$ then b is said to be the <u>order of a modulo c</u>, written $\mathrm{ord}_c\, a = b$.

<u>Theorem 2.14</u>. (Yamamoto, 1963) Let $q = 4t + 3$ be a prime divisor of v and let q^ℓ strictly divide v. Assume that any prime divisor p of n satisfies one of the conditions:

(i) order of p modulo q is even

(ii) order of p modulo q^ℓ is $q^{\ell-1}(q - 1)/2$,

(iii) $p = q$.

Then, if there exists a non-trivial v,k,λ - difference set, the diophantine equation

$$4n = x^2 + qy^2, \quad 0 \leq x, \quad 0 \leq y \leq vq^{-\ell}, \quad x + y \leq 2vq^{-\ell}$$

has a solution in integers x, y.

<u>Proof</u>. Let $w = q^\ell$, $\theta(\zeta_w) = \gamma$ and let σ be a generator of the Galois group of $K(\zeta_w)$ over the rational field; then σ takes ζ_w into ζ_w^s where s is a primitive root modulo w. If a prime p satisfies (ii) or (iii), it follows from Theorem 2.19 that all its prime ideal factors in $K(\zeta_w)$ are fixed by the automorphism σ^2. If p satisfies (i) then the order of p modulo w is also even; thus there exists an integer f such that $p^f \equiv -1 \pmod{w}$. This implies (Theorem 2.19) that the p-component of γ in $K(\zeta_w)$ is fixed under complex

conjugation and thus (from $\gamma\bar{\gamma} = n$) it must be the principal ideal (p^a) for some integer a. Thus in $K(\zeta_w)$ all the prime ideal factors of γ are fixed under σ^2. Hence the principal ideals (γ) and (γ^{σ^2}) are the same, i.e., $\gamma = \eta\gamma^{\sigma^2}$ where η is a unit of $K(\zeta_w)$. It follows from (2.20) and Theorem 2.20 that η thus must be a root of unity in $K(\zeta_w)$, i.e., $\eta = \varepsilon\, \zeta_w^j$ for some integer j where $\varepsilon = \pm 1$.

Let $N = \varphi(w) = q^{\ell-1}(q - 1)$, then N/2 is odd since $q \equiv 3\ (4)$. So $\varepsilon^{N/2} = \varepsilon$ and

$$1 = \eta^{1+\sigma^2+\dots+\sigma^{N-2}} = \varepsilon^{N/2}\, \zeta_w^{j(1+\sigma^2+\dots+\sigma^{N-2})} \tag{2.39}$$

shows that ε is a wth root of unity; thus $\varepsilon = +1$. If the difference set D is replaced by its shift D + u then $\eta = \gamma^{1-\sigma^2}$ is replaced by

$$(\zeta_w^u\, \gamma)^{1-\sigma^2} = \zeta_w^{u(1-\sigma^2)}\eta = \zeta_w^{(1-s^2)u+j} .$$

Thus there will be no loss of generality in assuming that $\eta = +1$ provided that it can be shown that

$$(1 - s^2)u \equiv - j \qquad \text{modulo } q^\ell \tag{2.40}$$

has a solution u. If $q \neq 3$ then (2.40) clearly does have a solution since then $1 - s^2$ is prime to q. When $q = 3$, the 3-component of $1 - s^2$ is 3 and the 3-component of $1 - s^N$ is 3^ℓ since s is a primitive root of 3^ℓ. Thus equation (2.39) [i.e., $j(1 - s^N)/(1 - s^2) \equiv 0$ modulo 3^ℓ] provides that 3 divides j, which of course guarantees a solution u for (2.40) in this case also. So, without loss of generality, $\eta = +1$ and thus γ is fixed by the automorphism σ^2. So γ is an algebraic integer of the quadratic subfield $K(\sqrt{-q})$ and $n = \gamma\bar{\gamma}$ is the norm of an integer of $K(\sqrt{-q})$. So $\gamma = a + b\omega$ where a, b are rational integers and $\omega = (-1 + \sqrt{-q})/2$; thus $4n = (2a - b)^2 + b^2q$. [Note that the complementary difference set D* has $\gamma^* = - a - b\omega$, that $\bar{\gamma} = a - b - b\omega$

and that $\bar{\gamma}* = b - a + b\omega$.] Thus by using $\bar{\gamma}$ or by replacing the difference set by its complement or by doing both it can be assumed that $a \geq 0$ and $b \geq 0$.

From (2.30) it follows that

$$\omega = \sum_{i=1}^{q-1} \psi(i)\, \zeta_q^i = \pm \sum_{i=1}^{q-1} \psi(i)\, \zeta_w^{iq^{\ell-1}} .$$

Here $\psi(i) = 1$ or 0 according as i is a quadratic residue or nonresidue of q, $\zeta_q = e^{2\pi i/q}$ and the sign $\pm$ depends upon the value of j in the relation $\zeta_w^{q^{\ell-1}} = \zeta_q^j$ (+ if j is a quadratic residue of q and - otherwise). Let $\theta(x) \equiv \Sigma_{i=0}^{w-1} B(i)\, x^i$ modulo $x^w - 1$, then the polynomial

$$f(x) = \sum_{i=0}^{w-1} B(i)x^i - \left(a \pm b \sum_{j=1}^{q-1} \psi(j)x^{jq^{\ell-1}} \right) = \sum_{i=0}^{w-1} C(i)x^i$$

is zero for $x = \zeta_w$; so apply Theorem 2.10 with $s = d = 1$ and $\alpha = 0$. This yields $\Delta(q^{\ell-1})\, C(i) = 0$ for all i. In particular, let $i = 0, q^{\ell-1}, \ldots, q^{\ell} - q^{\ell-1}$ then

$$B(0) - a = B(q^{\ell-1}r) \mp b\, \psi(r) \qquad r = 1, \ldots, q - 1$$

and since $\psi(1) = 1$, $\psi(q - 1) = 0$ it follows that $B(0) - a = B(q^{\ell-1}) \mp b = B(-q^{\ell-1})$. As $a \geq 0$, $b \geq 0$ and $0 \leq B(i) \leq vq^{-\ell}$ this implies that $0 \leq a, b \leq vq^{-\ell}$. Applying this same process to $\bar{\gamma}* = b - a + b\omega$ yields $|a - b| \leq vq^{-\ell}$. Combining these results establishes the theorem with $x = |2a - b|$ and $y = b$.

Yamamoto shows further that when $v = w = q^{\ell}$ the only non-trivial difference sets which satisfy the conditions of Theorem 2.14 are equivalent to the set D consisting of the quadratic residues of q or to its complementing set $D*$. Thus, in particular, Theorem 2.14 shows that no difference set exists with parameters

$v, k, \lambda = 239, 35, 5$ (a parameter set not ruled out by the previous results of this section).

Theorem 2.15. (Yamamoto, 1963) Let $q = 4t + 3$ and r be distinct odd prime divisors of v and let q^{ℓ}, r^m strictly divide v with $(\varphi(q^{\ell}), \varphi(r^m)) = 2$, where φ denotes Euler's function. Assume that any prime divisor p of n satisfies one of the conditions:

(i) $\operatorname{ord}_q p \equiv 0 \pmod 2$ and $\operatorname{ord}_r p \equiv 0 \pmod 2 \not\equiv 0 \pmod 4$,

(ii) order of p modulo q^{ℓ} is $\frac{1}{2}\varphi(q^{\ell})$ and order of p modulo r^m is $\varphi(r^m)$,

(iii) $p = q$ and order of p modulo r^m is $\varphi(r^m)$.

Then, if there exists a non-trivial v, k, λ - difference set, the equation

$$4n = x^2 + qy^2, \quad 0 \leq x, \quad 0 \leq y \leq 2vq^{-\ell}r^{-m}, \quad x + y \leq 4vq^{-\ell}r^{-m}$$

has a solution in integers x, y.

Note that, by Theorem 2.7, these hypothesis are only satisfied if n is a square. Thus as a non-existence test, applied after Theorem 2.7, it can be restricted to those cases. The integers 3, 6, 7, 12, 14 form a difference set for parameters 21, 5, 1. With $q = 7$, $r = 3$ the prime 2 satisfies condition (ii) and the equation has solution $x = 4$, $y = 0$. On the other hand, the parameters $v, k, \lambda = 306, 61, 12$ are not associated with any difference set since $p = 7$, $q^{\ell} = 9$, $r^m = 17$ satisfy (ii) of Theorem 2.15 and the diophantine equation has no solution.

Proof. Let $Q = q^{\ell}$, $R = r^m$, $w = q^{\ell}r^m$ and $\theta(\zeta_w) = \gamma$. Note that the Galois group of $K(\zeta_w)$ over the rational field is generated by two automorphisms σ and ρ where σ [respectively ρ] fixes every element of the field $K(\zeta_R)$ [respectively $K(\zeta_Q)$] and generates the Galois group of $K(\zeta_Q)$ [respectively $K(\zeta_R)$] over the rationals. If p is a prime divisor of n which satisfies (i), then there exists an integer f such that $p^f \equiv -1$ modulo w; thus (as in the previous proof) the p-component of the ideal (γ) is the principal ideal (p^a) for some integer a. Similarly, if p satisfies (iii) then the p-component of (γ) is (p^b) for some integer b (a consequence of Theorems 2.7 and 2.19). Finally, if p satisfies (ii) it follows from $(\varphi(q^\ell), \varphi(r^m)) = 2$ that the order of p modulo w is $\frac{1}{2}\varphi(w)$; thus there are two (Theorem 2.19) prime ideal divisors of (p) in $K(\zeta_W)$ and since $\operatorname{ord}_Q p = \frac{1}{2}\varphi(q^\ell)$ they originate in the quadratic subfield $K(\sqrt{-q})$.

Hence the ideal (γ) is fixed by ρ, fixed by σ^2 and originates in $K(\sqrt{-q})$. Thus $\gamma^{1-\sigma^2} = \eta$, $\gamma^{1-\rho} = \delta$ where η, δ are units in $K(\zeta_W)$; this implies (by 2.20) that $\eta\bar{\eta} = \delta\bar{\delta} = 1$ and that (Theorem 2.20) η, δ are roots of unity in $K(\zeta_W)$. Now $\eta^{1-\rho}$ is an r^mth root of unity and $\delta^{1-\sigma^2}$ is a q^ℓth root of unity. But both are $\gamma^{(1-\sigma^2)(1-\rho)}$; hence $\eta^{1-\rho} = \delta^{1-\sigma^2} = +1$. Thus η is a root of unity of $K(\zeta_Q)$ and δ is a root of unity in the subfield fixed by σ^2, i.e., in the field $K(\zeta_R, \sqrt{-q})$ of degree $2\varphi(r^m)$ over the rationals. Thus $\eta = \varepsilon\zeta_Q^j$ and $\delta = \varepsilon'\zeta_R^i$ for some integers i, j with $\varepsilon, \varepsilon' = \pm 1$ unless $q = 3$; if $q = 3$ then $\delta = \varepsilon'\zeta_3^a\zeta_R^i$ may occur for some integer a. [This last follows from the fact that $K(\sqrt{-q})$ $q \equiv 3(4)$ has no units except ± 1 unless $q = 3$.] By the same proof as above, see equation (2.39), it follows that $\varepsilon = +1$ and that by considering the shift $D + \mu r^m$ for a judicious choice of μ one may assume, without loss of generality, that $\eta = +1$. [Shifting D by a multiple of r^m has no effect on δ.] Shifting again by a multiple of q^ℓ does not disturb η but allows δ to take the shape ε' or $\varepsilon'\zeta_3^a$ for $q = 3$.

First consider the case $q \neq 3$. Since $\gamma^{1-\sigma^2} = \eta = 1$ it follows that γ is an integer of $K(\zeta_R, \sqrt{-q})$ and since $\gamma^{1-\rho} = \varepsilon' = \pm 1$ it follows that γ^2 is fixed by ρ, i.e., γ^2 is an integer of $K(\sqrt{-q})$. Also $\gamma^{1-\rho^2} = 1$, so γ belongs to $K(\sqrt{r^*}, \zeta_Q)$ where $r^* = (-1)^{(r-1)/2}r$ as before. That is, γ is an integer of the biquadratic field $\Omega = K(\sqrt{-q}, \sqrt{r^*})$. If $\varepsilon' = -1$ [i.e. if γ does not belong to $K(\sqrt{-q})$], then γ necessarily generates Ω over $K(\sqrt{-q})$. In this case, for some integer a in $K(\sqrt{-q})$, $x^2 - a = 0$ is the irreducible polynomial satisfied by γ over $K(\sqrt{-q})$ and thus [Mann (1955) Chapter 12] the relative different of γ over $K(\sqrt{-q})$ is 2γ. Since r is odd, $\gamma\bar{\gamma} = n$ and $(r,n) = 1$ is implied by conditions (i), (ii), (iii), it follows that the discriminant of Ω is prime to r; a contradiction. Thus $\varepsilon' = 1$ and γ is an algebraic integer of $K(\sqrt{-q})$.

Let $q = 3$. Since γ satisfies $\gamma^{1-\sigma^2} = 1$ it follows that γ, and thus γ^2, belongs to $K(\zeta_R, \sqrt{-3})$. Since $\gamma^{2(1-\rho)} = \zeta_3^{2a}$ it follows that either γ^2 belongs to $K(\sqrt{-3}) = K(\zeta_3)$ [here $a \equiv 0\ (3)$] or γ^2 is cubic over $K(\zeta_3)$ [see van der Waerden (1949) p. 171 for this fact]. Thus γ^2 determines a cubic extension Ω of $K(\sqrt{-3})$. Now r divides the discriminant of Ω since Ω is a subfield of $K(\zeta_R, \sqrt{-3})$ properly containing $K(\sqrt{-3})$. On the other hand γ^2 satisfies an irreducible equation $x^3 - b = 0$ for some integer b of $K(\sqrt{-3})$; thus its relative different is $3\gamma^4$, which is prime to r since $(r,n) = 1$ and since $r \neq 3 = q$. Contradiction. Thus $a \equiv 0\ (3)$ and γ^2 belongs to $K(\sqrt{-3})$. So this case has been reduced to the conditions for $q \neq 3$ and, since 3 plays no essential role in that argument, it may be invoked here to show that $\varepsilon' = 1$ and that γ is an algebraic integer of $K(\sqrt{-3})$.

Thus in every case $\gamma = \theta(\zeta_w)$ is an integer of $K(\sqrt{-q})$. As in the previous theorem, this implies that [with $\gamma = a + b\omega$, $\omega = \Sigma_{e=1}^{q-1} \psi(e)\zeta_q^e$, $\psi(e) = 1$ or 0 according as e is a quadratic residue or nonresidue of q, $\theta(x) \equiv \Sigma_{i=0}^{w-1} B(i)\, x^i$ modulo $x^w - 1$]

$$f(x) = \sum_{i=0}^{w-1} B(i)x^i - \left(a \pm b \sum_{e=1}^{q-1} \psi(e)\, x^{er^m q^{\ell-1}} \right) = \sum_{i=1}^{w-1} C(i)x^i$$

has a zero at $x = \zeta_w$. This implies (Theorem 2.10 with $s = 2$, $d = 1$ and $\alpha = 0$) that $\Delta(wq^{-1})\, \Delta(wr^{-1})\, C(i) = 0$ for all i. That is,

$$C(i) - C(i + q^{\ell-1}r^m) = C(i + q^\ell r^{m-1}) - C(i + q^\ell r^{m-1} + q^{\ell-1}r^m) \qquad (2.41)$$

for all i. Combining equations (2.41) for $i = 0, q^{\ell-1}r^m, \ldots, (q-2)q^{\ell-1}r^m$ yields

$$B(0) - a - B(q^{\ell-1}r^m j) \mp b\psi(j) = B(q^{\ell}r^{m-1}) - B(q^{\ell}r^{m-1} + jq^{\ell-1}r^m) \qquad (2.42)$$

for $j = 1, \ldots, q-1$. When $j = 1$ and $j = q-1$ equation (2.42) becomes

$$B(0) - a - B(q^{\ell-1}r^m) \mp b = B(q^{\ell}r^{m-1}) - B(q^{\ell}r^{m-1} + q^{\ell-1}r^m) \qquad (2.43)$$

$$B(0) - a - B(-q^{\ell-1}r^m) \quad = B(q^{\ell}r^{m-1}) - B(q^{\ell}r^{m-1} - q^{\ell-1}r^m)\,. \qquad (2.44)$$

Now $0 \leq B(i) \leq v/w$ for all i. Thus equation (2.44) shows that $|a| \leq 2v/w$ and $|b| \leq 2v/w$ follows from the same reasoning applied to the equation which results from subtracting (2.44) from (2.43). Consideration of $\overline{\gamma}^* = b - a + b\omega$ and the resulting equation analogous to (2.44) produces $|b - a| \leq 2v/w$. Thus with $x = |2a - b|$, $y = |b|$ the equation $4n = x^2 + qy^2$ has a solution of the proper type and the theorem has been proved.

Yamamoto notes that when $v = w = q^{\ell}r^m$ the only difference sets satisfying all the conditions of this theorem have parameters $v,k,\lambda = 21, 5, 1$; thus they are all equivalent to the set $\{3,6,7,12,14\}$.

Theorem 2.16. (Yamamoto, 1963) Let $q = 4t + 3$ and $r = 4s + 1$ be prime divisors of v, let q be a quadratic non-residue of r and let $(\varphi(q^{\ell}), \varphi(r^m)) = 2$ where q^{ℓ}, r^m strictly divide v. Assume that any prime divisor p of n satisfies one of the conditions:

(i) $\operatorname{ord}_q p \equiv 0 \pmod 2$ and $\operatorname{ord}_r(p) \equiv 0 \pmod 2 \not\equiv 0 \pmod 4$)

(ii) order of p modulo q^{ℓ} is $\varphi(q^{\ell})$ and order of p modulo r^m is $\varphi(r^m)$.

Then, if there exists a non-trivial v,k,λ - difference set, the equation

$$4n = x^2 + qry^2, \quad 0 \leq x, \quad 0 \leq y \leq 2vq^{-\ell}r^{-m}, \quad x + y \leq 4vq^{-\ell}r^{-m}$$

has a solution in integers x, y.

Again, as a non-existence test, applied after Theorem 2.7, this can be restricted to those cases where n is a square. The integers $0,1,2,4,5,8,10$ form a difference set with parameters $v,k,\lambda = 15, 7, 3$. Here the prime 2 satisfies (ii) for $q = 3$ and $r = 5$ and the equation is satisfied by $x = y = 1$. On the other hand, the parameter set $v,k,\lambda = 286, 96, 32$ is ruled out by this test, since $q = 11$, $r = 13$, $p = 2$ satisfies (ii) and no solution to the diophantine equation exists.

Proof. As in the proof of Theorem 2.15 above, let $Q = q^{\ell}$, $R = r^m$, $w = q^{\ell}r^m$, $\theta(\zeta_w) = \gamma$ and let σ, ρ generate the Galois group of $K(\zeta_w)$ over the rational field, σ generating the Galois group of $K(\zeta_Q)$ and fixing $K(\zeta_R)$ while ρ generates the Galois group of $K(\zeta_R)$ and fixes $K(\zeta_Q)$. If p is a prime divisor of n which satisfies (i) then it follows, exactly as in Theorem 2.15, that the p-component of the ideal (γ) is the principle ideal (p^a) for some integer a. While if p satisfies (ii) it follows from $(\varphi(q^{\ell}),\varphi(r^m)) = 2$ that $\mathrm{ord}_w\, p = \frac{1}{2}\varphi(w)$; thus there are two (Theorem 2.19) prime ideal divisors of (p) in $K(\zeta_w)$ and since $\mathrm{ord}_Q\, p = \varphi(q^{\ell})$, $\mathrm{ord}_R\, p = \varphi(r^m)$ they do not arise in $K(\zeta_Q)$ or in $K(\zeta_R)$. Since [see Mann (1955), Chapter 13] they must arise in a quadratic subfield of $K(\zeta_w)$ it can only be $K(\sqrt{-qr}\,)$, the field fixed by $\sigma\rho$.

Hence the ideal (γ) is fixed by $\sigma\rho$ and originates in $K(\sqrt{-qr}\,)$. Thus $\gamma^{1-\sigma\rho} = \eta$ is a unit of $K(\zeta_w)$ and by (2.20) and Theorem 2.20 it must be a root of unity in this field, i.e., $\eta = \pm\, \zeta_w^j$ for some integer j. By replacing the difference set D by a suitable shift $D + \mu q^{\ell} + \mu' r^m$ it may be assumed that $\eta = \pm 1$. Since $(\theta(q^{\ell}),\varphi(r^m)) = 2$, both σ^2 and ρ^2 are powers of $\sigma\rho$. In fact they are both <u>even</u> powers of $\sigma\rho$ and so $\gamma^{1-\sigma^2} = \gamma^{1-\rho^2} = +1$. Thus γ is fixed by both σ^2 and ρ^2 and as such lies in the biquadratic field $\Omega = K(\sqrt{r}, \sqrt{-q}\,)$. Furthermore, γ^2 is fixed by $\sigma\rho$ and thus belongs to $K(\sqrt{-qr}\,)$. Suppose $\gamma^{\sigma\rho} = -\gamma$, then since γ is an integer of Ω and since $1, (1 + \sqrt{r})/2$, $(1 + \sqrt{-q})/2$, $(1 + \sqrt{r} + \sqrt{-q} + \sqrt{-qr})/4$ form an integral basis for Ω, it follows that $\gamma = (c\sqrt{-q} + d\sqrt{r}\,)/2$ for some rational integers c, d such that

$c \equiv d$ modulo 2. Thus $4n = 4\gamma\bar{\gamma} = c^2q + d^2r$, and since n is necessarily a square, this implies that the diophantine equation $x^2 - qy^2 - rz^2 = 0$ has a solution in integers $x,y,z = 2\sqrt{n}$, c, d not all zero. But Legendre's test (see II.B note 2) shows that this only happens when q is a quadratic residue of r, which contradicts the hypothesis. Thus $\gamma^{\sigma\rho} = \gamma$ i.e., γ is an algebraic integer of $K(\sqrt{-qr})$.

So $\gamma = a + b\omega$ for some rational integers a, b where $\omega = (-1 + \sqrt{-qr})/2$ is the Gaussian sum $\Sigma_{i=1}^{qr-1} \psi(i)\zeta_{qr}^{i}$. Here $\zeta_{qr} = e^{2\pi i/qr}$ and $\psi(i) = \left[\left(\frac{i}{qr}\right) + 1\right]/2$ where $\left(\frac{i}{qr}\right)$ is the Jacobi symbol (see Nagell, 1951); that is $\left(\frac{i}{qr}\right) = 0$ if $(i,qr) > 0$, $= 1$ if i is a quadratic residue of both q and r or if i is a quadratic non-residue of both q and r, $= -1$ otherwise. [This representation of ω is a consequence of the analoguous representations of $\sqrt{r}$, $\sqrt{-q}$ given by equation (2.30)]. Let $\theta(x) \equiv \Sigma_{i=1}^{w-1} B(i)\, x^i$ modulo $x^w - 1$ and let

$$f(x) = \sum_{i=0}^{w-1} 2B(i)x^i - \left(2a \pm 2b \sum_{j=1}^{qr-1} \psi(j)x^{jq^{\ell-1}r^{m-1}}\right) = \sum_{i=0}^{w-1} C(i)x^i$$

then $f(x)$ has a zero at $x = \zeta_w$. Apply Theorem 2.10 with $s = 2$, $d = 1$, $\alpha = 0$; thus $\Delta(wq^{-1})\,\Delta(wr^{-1})\,C(i) = 0$ for all i. That is, equation (2.41) holds and combining these equations for $i = 0, q^{\ell-1}r^m, \ldots, (q-2)q^{\ell-1}r^m$ yields

$$2B(0) - 2a - 2B(jq^{\ell-1}r^m) \pm b = 2B(q^{\ell}r^{m-1}) \mp b - 2B(q^{\ell}r^{m-1} + jq^{\ell-1}r^m) \pm 2b\,\psi(q + jr).$$

From which, for $j = 1$, it follows that $|a| \leq 2v/w$. Combining these equations for $j = 1, j'$ (j' a quadratic non-residue of q) provides, as in the proof of Theorem 2.15, that $|b| \leq 2v/w$ also. Further, the analogous equation (for $j = 1$) associated with $\bar{\gamma}^*$ yields $|a - b| \leq 2v/w$. Thus with $x = |2a - b|$, $y = |b|$ the equation $4n = x^2 + qry^2$ has a solution of the proper type and the theorem has been proved.

Yamamoto notes that in this theorem when $v = w$ then in fact $v = qr$ where q, r are twin primes (i.e., either $q = r + 2$ or $r = q + 2$) and the only non-trivial cyclic difference sets which occur are the so-called twin prime difference sets; these have parameters $v, k, \lambda = qr, (v - 1)/2, (v - 3)/4$. These difference sets are discussed further in section V.D. below.

Let p be a prime divisor of both v and n, then since $k^2 = \lambda v + n$, $n = k - \lambda$ it follows that p divides k and λ also. In fact if p^r <u>strictly</u> divides k the relation $k(k - 1) = \lambda(v - 1)$ shows that p^r <u>strictly</u> divides λ; so p^r divides n also. Suppose p^{r+s} is the p-component of n, $s \geq 0$. Then

(i) $r > s$ implies p^s <u>strictly</u> divides v

(ii) $r < s$ implies p^r <u>strictly</u> divides v.

Thus in these cases, with p^{α} the p-component of v, it follows that $\alpha < (r + s)/2$. Finally

(iii) $r = s$ implies p^r divides v.

Now, in this case, let $p = 2$. Then n is a square; thus $k = 2^r k_1$, $n = 2^{2r} n_1^2$ where k_1 and n_1 are odd. Further $k^2 - n = 2^{2r}(k_1^2 - n_1^2) = \lambda v$, which shows that 2^{r+3} divides v since $k_1^2 - n_1^2 \equiv 0$ modulo 8. The next result shows that, in fact, no cyclic difference set has such parameters. This quite specialized result has application to some of the difference sets of section IV.C. below, as well as elsewhere.

<u>Theorem 2.17.</u> (Turyn, 1965) No cyclic difference set may have parameters v, k, λ such that 2^{2t} $(t \geq 1)$ strictly divides n if 2^{t+1} divides v.

<u>Proof</u>. By (i), (ii) only case (iii) need be considered. Here 2^t <u>strictly</u> divides k and λ while 2^{t+3} divides v. Let 2^{t+s} $(s \geq 3)$ <u>strictly</u> divide v, let $w = 2^{t+s-j}$ $(0 \leq j \leq t + s)$ and let $\theta(x) \equiv \Sigma_{i=0}^{w-1} C^j(i)x^i$ modulo $x^w - 1$. From $k \equiv 0$ modulo 2^t and Theorem 2.19 it follows that $\theta(\zeta_w^r) \equiv 0$ modulo 2^t for all integers r. So, by Theorem 2.10, $\Delta(w/2)C^j(i) \equiv 0$ modulo 2^t; i.e.

$$C^j(i + w/2) \equiv C^j(i) \qquad \text{modulo } 2^t$$

and thus $C^{j+1}(i) = C^j(i + w/2) + C^j(i) \equiv 2C^j(i) \qquad \text{modulo } 2^t$

In particular

$$C^t(i) \equiv 2C^{t-1}(i) \equiv \cdots \equiv 2^t C^0(i) \equiv 0 \qquad \text{modulo } 2^t$$

for all i.

Let $Z(i) = C^t(i)/2^t$, then for $w = 2^s$

$$\left(\sum_{i=0}^{w-1} Z(i)\,\zeta_w^{mi}\right)\left(\sum_{i=0}^{w-1} Z(i)\,\zeta_w^{-mi}\right) = n_1^2 \qquad (2.45)$$

for all $m \not\equiv 0$ modulo w, where $n = 2^{2t} n_1^2$. When w divides m the right side of this equation becomes k_1^2 where the odd integer k_1 satisfies $k = 2^t k_1$. Summing (2.45) for $m = 0, 1, \ldots, w-1$ yields

$$k_1^2 + (2^s - 1)n_1^2 = 2^s \sum_{i=0}^{w-1} Z^2(i) + \sum_{i \neq j} Z(i)Z(j) \sum_{m=0}^{w-1} \zeta_w^{(i-j)m}.$$

Since this last sum is zero for all $i \neq j$ it follows that

$$\sum_{i=0}^{w-1} Z^2(i) = n_1^2 + \frac{k_1^2 - n_1^2}{2^s}.$$

Since 2^s strictly divides $k_1^2 - n_1^2$ [see case (iii) above] the sum $\Sigma_{i=0}^{w-1} Z^2(i)$ is thus even. On the other hand $\Sigma_{i=0}^{w-1} Z(i) = k_1$ is an odd sum of integers, contradiction. Thus no such difference sets exist, as was to be proved.

All parameter sets $v, k, \lambda = 4N^2, 2N^2 - N, N^2 - N$ for $N = 2m$ are shown not to process cyclic difference sets by this theorem. Among others, the case $N = 28$ is not excluded by previous results.

The following facts of algebraic number theory are appended for easy reference. Proofs may be found, for example, in Mann's book (1955).

Theorem 2.18. The field $K(\zeta_d)$ is a normal extension of the rationals of degree $\varphi(d)$. The conjugates of ζ_d are the $\varphi(d)$ numbers ζ_d^i where $(i,d) = 1$. The numbers $1, \zeta_d, \ldots, \zeta_d^{\varphi(d)-1}$ form an integral basis for $K(\zeta_d)$ over the rationals. If $(r,s) = 1$ the field $K(\zeta_{rs})$ is of degree $\varphi(r)$ over $K(\zeta_s)$ and any $\varphi(r)$ consecutive powers of ζ_r form an integral basis for $K(\zeta_{rs})$ over $K(\zeta_s)$.

Theorem 2.19. The prime ideal decomposition of the rational prime p [that is of the principal ideal (p)] in $K(\zeta_d)$ is given by

$$(p) = (1 - \zeta_d)^{\varphi(d)} \qquad \text{when} \qquad d = p^i \tag{2.46}$$

$$(p) = P_1 \ldots P_g \qquad \text{when} \qquad (d,p) = 1 \tag{2.47}$$

where the distinct prime ideals P_i are conjugates. Further, g is $\varphi(d)/f$ where f is the order of p modulo d. The field automorphism determined by the mapping $\zeta_d \to \zeta_d^p$ fixes each of these prime ideals P_i.

$$(p) = (P_1 \ldots P_g)^{\varphi(p^a)} \qquad \text{when} \qquad d = p^a w,\ (p,w) = 1 \tag{2.48}$$

where g is $\varphi(w)/f$ and f is the order of p modulo w. The ideals $P_1, \ldots, P_g$ of (2.47), (2.48) can be determined explicitly. For, if $f(x)$ denotes the irreducible equation satisfied by ζ_d over the rationals then, with the f_i irreducible modulo p,

$$f(x) \equiv (f_1(x) \ldots f_g(x))^{\varphi(p^a)} \qquad \text{modulo } p$$

and

$$P_i = (p, f_i(\zeta_d)) .$$

Theorem 2.20. Any algebraic integer which lies, together with all its conjugates, on the unit circle is a root of unity.

III. MULTIPLIERS AND CONSTRUCTIVE EXISTENCE TESTS

The mere fact that parameters v, k, λ have passed the existence tests of Chapter II does not insure the existence of an associated difference set. Of course the parameters may well belong to a known family of difference sets (see Chapter V for a discussion of these) in which case there is no problem. Often enough, however, the parameters neither belong to a known family nor are they excluded by the results of Chapter II. When this happens an attempt is usually made at constructing a difference set with the given parameters. The straightforward process of examining all sets of k elements taken from $0, 1, \ldots, v - 1$ soon becomes prohibitively large. Thus, more efficient methods have been devised; for the most part these methods depend upon the existence of a multiplier of the hypothetical difference set in question. This chapter is concerned first with general theorems about multipliers (multiplier results specific to certain types of difference sets are found in Chapters IV and V) and second with constructive existence tests, whether based on multipliers or not.

A. Multiplier Theorems

The question of whether every cyclic difference set necessarily possesses a non-trivial multiplier t (i.e., $t \not\equiv 1$ modulo v) is open. Thus, given parameters v, k, λ, one cannot in general assume the existence of a non-trivial multiplier. However, under certain circumstances, multipliers can be shown to exist - the main result of this type is due to Hall (1956):

Theorem 3.1. Let D be a v, k, λ - difference set and let n_0 be a divisor of n, where $(n_0,v) = 1$ and $n_0 > \lambda$. If, for every prime p dividing n_0, there is an integer j_p such that

$$p^{j_p} \equiv t \qquad (\text{modulo } v)$$

then t is a multiplier of D.

Proof. From

$$\theta(x)\theta(x^{-1}) \equiv n + \lambda(1 + x + \cdots + x^{v-1}) \qquad (\text{mod } x^v - 1)$$

comes the factorization

$$\theta(x)\theta(x^{-1}) \equiv n = n_0 n_1 \qquad \text{modulo} \quad f_i(x) \tag{3.1}$$

when $f_i(x)$ is any one of the distinct irreducible factors of $T(x) = 1 + x + \cdots + x^{v-1}$ over the rational field K. Let $\zeta = e^{2\pi ij/v}$ be that root of $f_i(x)$ for which j is least positive; then congruence 3.1 yields the factorization $\theta(\zeta)\theta(\zeta^{-1}) = n_0 n_1$. In the cyclotomic field $K(\zeta)$, the algebraic integers $1, \zeta, \ldots, \zeta^{u-1}$ form an integral basis [here u is the degree of $f_i(x)$] and the algebraic integers of $K(\zeta)$ can be associated with the residue classes of polynomials with rational integral coefficients modulo $f_i(x)$ [see, for example, Mann (1955) Theorem 8.6 for a proof of these facts].

Now if P is any prime ideal divisor of the rational prime p and if $(p,v) = 1$, then the automorphism α of the field $K(\zeta)$ determined by $\zeta \to \zeta^p$, leaves P fixed (i.e., $P^\alpha = P$). [See Mann (1955) Theorem 8.1.] Thus $\zeta \to \zeta^t$ determines an automorphism of $K(\zeta)$ which fixes all the prime ideal divisors of n_0. Hence n_0 divides $\theta(\zeta)\theta(\zeta^{-t})$ as well as $\theta(\zeta)\theta(\zeta^{-1})$, so

$$\theta(x)\theta(x^{-t}) \equiv n_0 S_i(x) \qquad \text{modulo} \quad f_i(x) \tag{3.2}$$

where $S_i(x)$ has rational integral coefficients. There is such a congruence for each irreducible factor $f_i(x)$ of $T(x)$. A similar congruence modulo $T(x)$ is desired. Suppose

$$\theta(x)\theta(x^{-t}) = n_0 R_j(x) + A(x)F_j(x) \tag{3.3}$$

where $F_j(x) = f_1(x)f_2(x) \dots f_j(x)$ and $A(x)$ as well as $R_j(x)$ are polynomials with rational integral coefficients. This is immediate for $j = 1$ from congruence 3.2. Assume equation (3.3) holds for j and consider $j + 1$. Now $f_{j+1}(x)$ and $F_j(x)$ have no common factors. Thus their resultant z [van der Waerden (1949) p. 83-87 establishes all the resultant theory needed here] is a non-zero rational integer, and there exist integral polynomials $C(x)$ and $D(x)$ such that

$$C(x)F_j(x) + D(x)f_{j+1}(x) = z \,. \tag{3.4}$$

But z can be expressed as a product of factors $\alpha - \beta$ where α is a root of $F_j(x)$ and β is a root of $f_{j+1}(x)$. Thus α and β are different v^{th} roots of unity, and if η is a primitive v^{th} roots of unity, $\alpha - \beta = \eta^y(\eta^\ell - 1)$ for appropriate exponents y, ℓ. Now η^y is unit of $K(\eta)$ and $\eta^\ell - 1$ is a root of

$$\frac{[(x + 1)^v - 1]}{x} = x^{v-1} + \dots + v$$

thus $\eta^\ell - 1$ divides v. Hence z will divide an appropriate power of v, and, since $(n_0, v) = 1$ has been assumed, then also $(n_0, z) = 1$. From congruence 3.2 with $i = j + 1$ follows

$$C(x)F_j(x)\theta(x)\theta(x^{-t}) = n_0 C(x)F_j(x)S_{j+1}(x) + C(x)B(x)F_{j+1}(x) \,. \tag{3.5}$$

Multiplying congruence 3.3 by $D(x)f_{j+1}(x)$ and adding the result to equation (3.5) yields [by (3.4)]

$$z\theta(x)\theta(x^{-t}) = n_0 S(x) + G(x)F_{j+1}(x) \,.$$

This can be combined with the trivial relation

$$n_0\theta(x)\theta(x^{-t}) = n_0 H(x)$$

to yield (since n_0 and z are relatively prime)

$$\theta(x)\theta(x^{-t}) = n_0 R_{j+1}(x) + A(x)F_{j+1}(x) .$$

Continuing in this manner provides the desired equation

$$\theta(x)\theta(x^{-t}) = n_0 R(x) + A(x)T(x) . \qquad (3.6)$$

Since $A(x)T(x) \equiv A(1)T(x)$ modulo $x^v - 1$, equation (3.6) yields

$$\theta(x)\theta(x^{-t}) \equiv n_0 R(x) + A(1)T(x) \qquad (\text{mod } x^v - 1) .$$

Let $x = 1$ in this congruence, then $k^2 = n_0 R(1) + vA(1)$. Since $k^2 = n + \lambda v$ in any difference set and $(n_0, v) = 1$ this implies $A(1) \equiv \lambda$ modulo n_0. Thus, by altering $R(x)$ if necessary,

$$\theta(x)\theta(x^{-t}) \equiv n_0 R(x) + \lambda T(x) \qquad (\text{mod } x^v - 1) . \qquad (3.7)$$

Now every coefficient on the left side of this congruence is non-negative and since $n_0 > \lambda$ all the coefficients of $R(x)$ are non-negative also. Further with $x = 1$ this congruence provides $k^2 = n_0 R(1) + \lambda v$; thus $R(1) = n_1$.

Since $(t,v) = 1$,

$$\theta(x)\theta(x^{-1}) \equiv \theta(x^t)\theta(x^{-t}) \equiv n_0 n_1 + \lambda T(x) \qquad (\text{mod } x^v - 1)$$

thus $\theta(x^t)\theta(x^{-t})\theta(x)\theta(x^{-1}) \equiv [n_0 n_1 + \lambda T(x)]^2$, while (3.7) gives

$$[n_0 R(x) + \lambda T(x)]\,[n_0 R(x^{-1}) + \lambda T(x)] \qquad (\text{mod } x^v - 1)$$

for this same product. As $R(x)T(x) \equiv R(1)T(x) \equiv n_1 T(x)$ modulo $x^v - 1$, a comparison of these two results yields

$$R(x)R(x^{-1}) \equiv n_1^2 \qquad (\text{mod } x^v - 1).$$

This implies [since $R(x)$ has non-negative coefficients] that $R(x)$ has only a single non-zero term, i.e., $R(x) \equiv n_1 x^{-s}$ (mod $x^v - 1$) for some integer s. Thus congruence 3.7 implies

$$\theta(x^{-1})\theta(x^t) \equiv nx^s + \lambda T(x) \qquad (\text{mod } x^v - 1).$$

Multiplying this last congruence by $\theta(x)$ and simplifying yields

$$\theta(x^t) \equiv x^s\theta(x) \qquad (\text{mod } x^v - 1)$$

i.e., t is a multiplier of this difference set as was to be proved.

This theorem was first established for cyclic projective planes (i.e., $\lambda = 1$ only) by Hall (1947) and later extended (Hall and Ryser, 1951) to the case of general λ in the form stated in I.E. above. Theorem 3.1 represents a further generalization. The conditions $n_0 > \lambda$ and $(n_0, v) = 1$ are superfluous in <u>every known</u> case. For, no cyclic difference sets are known with $(n,v) > 1$ and every prime divisor p of n is a multiplier of every known cyclic difference set.

Morris Newman (1963) extended this result slightly by showing that the odd prime p is always a multiplier whenever $n = 2p$ and $(7p, v) = 1$. Turyn (1964) generalized Newman's result.

<u>Theorem 3.1A</u> (Turyn) Let $n = 2n_0$ with n_0 odd and prime to v. Suppose that for every prime p dividing n_0 there is an integer j_p such that

$$p^{j_p} \equiv t \qquad \text{modulo } v.$$

Then t is a multiplier of every difference set with these parameters, provided merely that t is a quadratic residue of 7 whenever 7 is a divisor of v.

Note that in particular this eliminates the condition $(7,v) = 1$ from Newman's result. For if $n = 2p^a$ (a odd), the assumption that $t \equiv p^{j_p}$ is not a quadratic residue of 7 implies that $t^3 \equiv p^{3j_p} \equiv -1 \pmod 7$. So Theorem 2.7 provides the contradiction that a is necessarily even.

Mann and Zaremba (1969) investigate the situation when 7 divides v and t is not a quadratic residue of 7. However they do not resolve it completely. In particular, they find no case where t is not a multiplier.

On occasion it is possible to establish the existence of a w-multiplier for some divisor w of v even when it is not possible to show the existence of a multiplier. Frequently this is of importance in constructive existence tests. [Section III.C. contains an example of such a case.] By retracing the proof above under the conditions $(n_0,w) = 1$, $n_0 > \lambda v/w$, $p^{j_p} \equiv t \pmod w$ it can be established that t is a w-multiplier. However the condition $n_0 > \lambda v/w$ is rarely satisfied; thus this is not of much use. The following related result is more useful:

Theorem 3.2. Let w divide v and suppose that for every prime p dividing n there is an integer j_p such that

$$p^{j_p} \equiv t \qquad (\text{modulo } w)$$

with $(t,w) = 1$. Then t is a w-multiplier of every difference set with these parameters n, v.

Of course the proof of this theorem is almost identical to that of Theorem 3.1 above. The congruence analogous to (3.7) above, being

$$\theta(x)\theta(x^{-t}) \equiv nR(x) + \frac{\lambda v}{w}(1 + x + \cdots + x^{w-1}) \qquad (\text{mod } x^w - 1) \qquad (3.8)$$

with $R(1) = 1$. From which $R(x)R(x^{-1}) \equiv 1 \pmod{x^w - 1}$ is deduced in the manner above. But this implies that $R(x) = \pm x^{-s}$ for some s and since $R(1) = 1$, it must be that $R(x) = x^{-s}$. From which point on the proof concludes as for Theorem 3.1.

A surprisingly important fact about cyclic difference sets is that:

Theorem 3.3. Minus one is never a multiplier of a non-trivial cyclic difference set.

This fact was known for several years prior to any publication of its proof. This accounts for the anomaly that it is often referred to in publications which predate the papers [Johnsen (1964), Brualdi (1965) and Yates (1967)] containing proofs.

Proof. The result (proved in section III.B. below) that any multiplier of a difference set D fixes at least one shift $D + s$ of the difference set is used. Assume that -1 is a multiplier and that D is fixed by -1, i.e., that x belongs to D only if $-x$ $(= v - x)$ does also. Consider the differences $x - y \equiv d$ and $(-y) - (-x) \equiv d$ for x, y in D; they are distinct representations of d unless $x = -y$. Hence, if $0 \neq d \not\equiv 2x$ for some x in D (such d must exist if the difference set is non-trivial), then d occurs an even number of times as the difference of elements of D. Thus λ is necessarily even.

If $d \equiv 2x$ for some x in D, then $x - (-x) \equiv d$ occurs provided $0 \neq x \neq v/2$. So d can appear an even number of times as a difference of elements of D only if there exists a $y \neq x$ (y in D) such that $y - (-y) \equiv d$ also. Thus $2x \equiv 2y$, which can only happen for even v. Whence $x \equiv y + v/2$. Thus, for every x in D $(0 \neq x \neq v/2)$ the element $y \equiv x - v/2$ also belongs to D. Thus the difference $v/2$ is represented at least $k - 1$ times. Hence $\lambda \geq k - 1$ and the difference set is trivial. So, only trivial cyclic difference sets have multiplier -1 as was to be shown.

It is, of course, not necessary that a multiplier t divide n. However, it has been shown (Mann, 1952) that 2 is a multiplier of a non-trivial difference set only when n is even.

B. Difference Sets Fixed by a Multiplier

The use of a multiplier t for constructing a difference set is greatly facilitated by the assumption that the set is fixed by the multiplier (i.e., that $tD \equiv D$ modulo v). As mentioned in section I.E. every multiplier t of a difference set determines an automorphism of the associated block design. That is, if A is the incidence matrix of the block design, then the multiplier t determines permutation matrices P and Q (Q takes column x into column tx modulo v) such that

$$PAQ = A \qquad \text{and thus} \qquad A^{-1}PA = Q^{-1} \tag{3.9}$$

since (see section I.C.) A is non-singular for non-trivial designs. Hence (by well known facts of linear algebra)

$$\mathrm{Tr}(P) = \mathrm{Tr}(A^{-1}PA) = \mathrm{Tr}(Q^{-1}) = \mathrm{Tr}(Q) \tag{3.10}$$

where $\mathrm{Tr}(X)$ denotes the trace of the matrix X (i.e., the sum of its diagonal elements). But $\mathrm{Tr}(P)$ is the number of blocks (or shifts of the difference set) fixed by the multiplier, whereas $\mathrm{Tr}(Q)$ is the number of objects fixed by the multiplier. So, by equation (3.10), the number of shifts fixed by the multiplier is the number of solutions of $tx \equiv x$ modulo v, which is $(t - 1, v) = d$. Hence

Theorem 3.4. Given a difference set D with multiplier t, there exists exactly $(t - 1, v) = d$ shifts fixed by the multiplier. In fact if E is a shift fixed by t, then t also fixes the d shifts $E + j(v/d)$, for $j = 0, 1, \ldots, d-1$.

[Similarly, if t is a w-multiplier for some divisor w of v then t necessarily fixes exactly $(t - 1, w) = \delta$ shifts of D modulo w. If E modulo w is one such shift, then all others are of the form $E + j(w/\delta)$ modulo w, for $j = 0, 1, \ldots, \delta - 1$.]

Suppose t_1, t_2 are both multipliers of the same difference set D and suppose that D is fixed by t_1. Then $t_1(t_2D) = t_2(t_1D) = t_2D$; that is, t_2D is also fixed by t_1. So

Theorem 3.5. If t_1, t_2 are multipliers of the same difference set then t_2

permutes the shifts fixed by t_1.

Thus if any multiplier t fixes only one shift of the difference set then that shift is fixed by all multipliers. Hence

Theorem 3.6. (1) If there exists a multiplier t such that $(t - 1, v) = 1$, then exactly one shift of the difference set is fixed by all multipliers. (2) If $(k,v) = 1$ then there exists at least one shift fixed by all multipliers.

Part (2) follows because $(k,v) = 1$ insures that exactly one shift E of the difference set has $e_1 + e_2 + \cdots + e_k \equiv 0 \pmod v$. Such a shift (being unique) is certainly fixed by any multiplier.

Some additional fixed shift results which apply only to the case $\lambda = 1$ are contained in section IV.A. The fact that every multiplier fixes at least one shift of the difference set was shown by McFarland and Mann (1965). Part (2) of Theorem 3.6 is due to J. Jans and the remainder of this section restates for arbitrary λ results of Hall (1947).

C. Multipliers and Diophantine Equations

Suppose it is known (by Theorem 3.1 or otherwise) that a hypothetical difference set has a multiplier t. Then (Theorem 3.4) some shift of this difference set is fixed by the multiplier. Thus, there is no loss of generality in assuming that the difference set is a union of sets $\{a, ta, \ldots, t^{m-1}a\}$, where $t^m a \equiv a$ modulo v. So

Lemma 3.7. If a v, k, λ difference set with multiplier t exists, some union of sets $\{a, ta, \ldots, t^{m-1}a\}$, where $t^m a \equiv a \pmod v$, has exactly k distinct elements and forms a difference set with parameters v, k, λ.

The sets $\{a, ta, \ldots, t^{m-1}a\}$ are often called blocks fixed by the multiplier t. The number of distinct elements in each of these fixed blocks is always a divisor of the order of t modulo v (this follows, for example, from Theorem 60 of Nagell, 1951) and is in fact always equal to that order unless $(a,v) > 1$. Thus when v is prime all the blocks (save $\{0\}$) fixed by a multiplier contain the same number of distinct elements m (= the order of t modulo v). In this case a difference set can exist only if $k = jm$ or $jm + 1$, for some integer j.

Lemma 3.7 is not always easy to apply (for sometimes there are many unions of these blocks having k distinct elements) however frequently it provides significant information almost trivially. Consider the parameters $v, k, \lambda = 151, 51, 17$. By Theorem 3.1 above, $t = 76$ [$\equiv 2^{14} \equiv (17)^{35}$ modulo 151] is a multiplier. Now 151 is prime and m (= order of 76 modulo 151) is 15; so a difference set could only exist if $k = 51$ were congruent to 0 or 1 modulo 15, contradiction. For the parameter values $21, 5, 1$, Theorem 3.1 establishes that 2 is a multiplier; the fixed blocks are $\{0\}$, $\{3, 6, 12\}$, $\{9, 15, 18\}$, $\{7, 14\}$ and two with 6 elements each. Thus the only difference set candidates are $\{3, 6, 7, 12, 14\}$ and $\{7, 9, 14, 15, 18\}$. These are equivalent (see section I.B. for this definition) difference sets, as is easily verified. [This difference set is the only non-trivial one known in which every residue d_i has a common factor with v.]

If a difference set exists, then its Hall-polynomial $\theta(x) = x^{d_1} + \cdots + x^{d_k}$ satisfies the congruence

$$\theta(x)\theta(x^{-1}) \equiv n + \lambda(1 + x + \cdots + x^{v-1}) \qquad (\text{mod } x^v - 1). \tag{3.11}$$

Thus, for any divisor w of v, one has

$$\theta(x) \equiv b_0 + b_1 x + \cdots + b_{w-1}x^{w-1} \qquad (\text{mod } x^w - 1) \tag{3.12}$$

$$\theta(x)\theta(x^{-1}) \equiv n + \frac{\lambda v}{w}(1 + x + \cdots + x^{w-1}) \qquad (\text{mod } x^w - 1) \tag{3.13}$$

where $b_i (\leq v/w)$ is the number of d_j in D satisfying $d_j \equiv i \bmod w$. This yields (comparing coefficients in congruence 3.13).

Lemma 3.8. If a difference set exists, then, for every divisor w of v, there exists integers b_i $(i = 0,\ldots,w - 1)$ satisfying the diophantine equations

$$\sum_{i=0}^{w-1} b_i = k, \quad \sum_{i=0}^{w-1} b_i^2 = n + \frac{\lambda v}{w}, \qquad 0 \leq b_i \leq v/w \tag{3.14}$$

and

$$\sum_{i=0}^{w-1} b_i b_{i-j} = \lambda v/w \tag{3.15}$$

for $j = 1,\ldots,w-1$. (Here the subscript $i-j$ is taken modulo w.)

An example of the application of Lemma 3.8 is provided by the parameters $v, k, \lambda = 70, 24, 8$. Here 2 is a w-multiplier with $w = 35$ (Theorem 3.2) and thus also for $w = 5, 7$. Consider a shift of $\theta(x)$ modulo $x^{35} - 1$ which is fixed under the multiplier 2. The residues modulo 35 break into fixed sets which impart certain restrictions on $\theta(x)$ modulo $x^5 - 1$ and modulo $x^7 - 1$.

Mod 35	Mod 7	Mod 5
0	0	0
1,2,4,8,16,32,29,23,11,22,9,18	4(1,2,4)	3(1,2,3,4)
3,6,12,24,13,26,17,34,33,31,27,19	4(3,6,5)	3(1,2,3,4)
5,10,20	3,6,5	3(0)
15,30,25	1,2,4	3(0)
7,14,28,21	4(0)	1,2,3,4

Thus

$$\theta(x) \equiv c_0 + c_1(x + x^2 + \cdots + x^{18}) + c_3(x^3 + \cdots + x^{19}) + c_5(x^5 + x^{10} + x^{20})$$

$$+ c_{15}(x^{15} + x^{30} + x^{25}) + c_7(x^7 + x^{14} + x^{28} + x^{21}) \qquad (\text{mod } x^{35} - 1)$$

where c_i $(0 \leq c_i \leq 2)$ is the coefficient of x^i. Further

$$\theta(x) \equiv a_0 + a_1(x + x^2 + x^4) + a_3(x^3 + x^6 + x^5) \qquad (\text{mod } x^7 - 1)$$

$$\theta(x) \equiv b_0 + b_1(x + x^2 + x^3 + x^4) \qquad (\text{mod } x^5 - 1)$$

where $0 \leq a_i \leq 10$ and $0 \leq b_i \leq 14$. For $w = 5$ equation (3.14) becomes $b_0 + 4b_1 = 24$, $b_0^2 + 4b_1^2 = 128$. Then $b_0 = 8$, $b_1 = 4$ is the unique solution. This implies (from the residue table above) that $c_7 = 1$ and hence that $a_0 \geq 4$. For $w = 7$ equation (3.14) becomes $a_0 + 3(a_1 + a_3) = 24$, $a_0^2 + 3(a_1^2 + a_3^2) = 96$ which has only two solutions with $a_0 \geq 4$; these are $a_0, a_1, a_3 = 6, 2, 4$ or $6, 4, 2$. Thus $a_0 = 6$ and hence $c_0 = 2$. When $a_1 = 2$, $a_3 = 4$ then $c_{15} = 2$, $c_3 = 1$, $c_1 = c_5 = 0$. If $a_1 = 4$, $a_3 = 2$ then $c_5 = 2$, $c_1 = 1$, $c_3 = c_{15} = 0$. Since $x \to x^3$ transforms this second solution into the first (and takes the difference set into an equivalent one) it is only necessary to consider one solution. Thus, without loss of generality, modulo $x^{35} - 1$

$$\theta(x) \equiv 2(1 + x^5 + x^{10} + x^{20}) + x + x^2 + x^4 + x^7 + x^8 + x^9 + x^{11} + x^{14}$$
$$+ x^{16} + x^{18} + x^{21} + x^{22} + x^{23} + x^{28} + x^{29} + x^{32}. \qquad (3.16)$$

It can be seen (after some searching) that no difference set polynomial $\theta(x)$ can satisfy congruence 3.16. Thus no difference set with parameters $v, k, \lambda = 70, 24, 8$ exists.

D. Polynomial Congruences

If a difference set exists then, for every divisor w of v, there must exist polynomials $\theta_w(x)$, $\theta_{[w]}(x)$ with rational integral coefficients such that

$$\theta(x) \equiv \theta_{[w]}(x) \qquad \text{modulo } x^w - 1 \qquad (3.17)$$

$$\theta(x) \equiv \theta_w(x) \qquad \text{modulo} \quad f_w(x) \tag{3.18}$$

where $f_w(x)$ is the irreducible polynomial satisfied by the primitive wth roots of unity over the rational field. Furthermore, the coefficients of $\theta_{[w]}(x)$ are non-negative. Conversely (by the Chinese Remainder Theorem), the set of polynomials $\{\theta_w(x),$ all w's dividing $v\}$ uniquely determines $\theta(x)$ modulo $x^v - 1$. In fact,

$$\theta(x) \equiv \frac{1}{v} \sum_{w|v} \theta_w(x)\, B_{v,w}(x) \qquad (\text{mod} \quad x^v - 1) \tag{3.19}$$

where

$$B_{v,w}(x) = \sum_{r|w} \mu\left(\frac{w}{r}\right) r \frac{x^v - 1}{x^r - 1} \tag{3.20}$$

and μ is the Möbius function. Similarly

$$\theta_{[w]}(x) \equiv \frac{1}{w} \sum_{d|w} \theta_d(x)\, B_{w,d}(x) \qquad (\text{mod} \quad x^w - 1) \tag{3.21}$$

Proof. By the Chinese Remainder Theorem, congruence 3.19 can be established by merely verifying the conditions imposed by congruence 3.18. That is, by verifying that

$$\frac{1}{v} \sum_{w|v} \theta_w(x)\, B_{v,w}(x) \equiv \theta_r(x) \qquad \text{modulo} \quad f_r(x) \tag{3.22}$$

for an arbitrary divisor r of v.

[To facilitate this as well as for future reference, recall that if $f(x)$, $g(x)$, $h(x)$ are integral polynomials and $f(x)$ is irreducible over the rational field, then $g(x) \equiv h(x)$ modulo $f(x)$ if and only if $g(r) = h(r)$ where r is any root of $f(x)$. Further recall that

$$x^m - 1 = \prod_{d|m} f_d(x) \quad \text{and} \quad f_d(x) = \prod_{h|d} (x^h - 1)^{\mu(d/h)} \tag{3.23}$$

where μ is the Möbius function.]

Now if r and s are divisors of m, then

$$\frac{x^m - 1}{x^s - 1} \equiv \begin{cases} m/s & \text{modulo } f_r(x) \text{ when } r \text{ divides } s \\ 0 & \text{modulo } f_r(x) \text{ otherwise} \end{cases}$$

This follows from (3.23) for r's <u>not</u> dividing s and can be seen by evaluating the left side at $e^{2\pi i/r}$ when r divides s. Thus, when d divides m

$$B_{m,d}(x) \equiv \sum_{r|s,\ s|d} \mu\left(\frac{d}{s}\right) s\,\frac{m}{s} \equiv m \sum_{r|s,\ s|d} \mu\left(\frac{d}{s}\right) \qquad \text{modulo } f_r(x)$$

$$B_{m,d}(x) \equiv \begin{cases} m & \text{modulo } f_d(x) \\ 0 & \text{modulo } f_r(x) \quad \text{for } r \neq d \end{cases} \tag{3.24}$$

by the standard property of the Möbius function. From this congruence 3.22 is immediate and so congruence 3.19 is established. For future use, note that when d divides m

$$B_{m,d}(x) \equiv \frac{x^m - 1}{f_d(x)} \; [x \, f_d'(x)] \qquad (\text{mod } x^m - 1) \qquad (3.25)$$

where (') denotes derivation with respect to x. This congruence may be verified by substituting the various m^{th} roots of unity for x and using congruence 3.24. The only difficulty arises when x is a primitive d^{th} root of unity; but l'Hospital's rule then shows that the right side is indeed m as desired.

A constructive existence test procedure can be based on congruence 3.21. If this congruence is used to construct $\theta_{[w]}(x)$, it requires the knowledge of integral polynomials $\theta_d(x)$ for all divisors d of w such that $\theta(x) \equiv \theta_d(x)$ modulo $f_d(x)$. Neglecting for the moment the problem of finding an exhaustive list of candidates for $\theta_d(x)$, note that given a set of integral polynomials $\{\theta_d\}$ (i.e., one θ_d for each divisor d of w) there is no guarantee that the polynomial computed from them by congruence 3.21 will have integral coefficients, much less that these coefficients will be non-negative. Of course, if these coefficients are not non-negative integers, then there is no difference set corresponding to this set of $\theta_d(x)$'s. Furthermore, if congruence 3.21 is being used to construct $\theta_{[w]}(x)$ for some divisor w of v, it is quite reasonable to expect that integral polynomials $\theta_{[d]}(x)$ (for all divisors d of w) will have been constructed previously. Thus, the assumption that they are known is not restrictive. This assumption allows the use of the following lemma, which helps screen the $\theta_d(x)$ candidates by imposing the condition that $\theta_{[w]}(x)$ have integral coefficients.

Lemma 3.9. Let w be a positive integer and suppose that, for each divisor d of w, an integral polynomial $\theta_d(x)$ is given. Further, assume that an integral polynomial $\theta_{[w/p]}(x)$ is known, for each prime p dividing w, and that these are consistent with the given $\theta_d(x)$'s, i.e., assume that

$$\theta_{[w/p]}(x) \equiv \theta_d(x) \qquad \text{modulo } f_d(x)$$

for all divisors d of w/p. Then, necessary and sufficient conditions for the existence of an integral polynomial $\theta_{[w]}(x)$, such that

$$\theta_{[w]}(x) \equiv \theta_d(x) \qquad \text{modulo} \quad f_d(x) \tag{3.26}$$

for all divisors d of w, are that

$$\theta_w(x) \equiv \theta_{[w/p]}(x) \qquad \text{mod}(p, f_{w_1}^{p^{a-1}}(x)) \tag{3.27}$$

for all primes p dividing w. Here $w = p^a w_1$ with p prime to w_1.

Proof. Let $\theta_{[w]}(x)$ be an integral polynomial satisfying congruence 3.26. By the Chinese Remainder Theorem

$$\theta_{[w]}(x) \equiv \theta_{[w/p]}(x) \qquad \text{mod}(x^{w/p} - 1)$$

so

$$\theta_{[w/p]}(x) \equiv \theta_w(x) \qquad \text{mod}[x^{w/p} - 1, \ f_w(x)] .$$

Thus

$$\theta_{[w/p]}(x) \equiv \theta_w(x) \qquad \text{mod}[p, f_{w_1}^{p^{a-1}}(x)]$$

i.e., congruence 3.27 has been established.

Conversely, assume $\theta_{[w]}(x)$ is the polynomial provided by the Chinese Remainder Theorem from the given $\theta_d(x)$'s (it may not have integral coefficients), then by congruence 3.24

$$[\theta_{[w]}(x) - \theta_w(x)] \, B_{w,w}(x) \equiv 0 \qquad (\text{mod } x^w - 1)$$

$$\sum_{s|w} \mu\left(\frac{w}{s}\right) s[\theta_{[w]}(x) - \theta_w(x)] \frac{x^w - 1}{x^s - 1} \equiv 0 \qquad (\text{mod } x^w - 1) . \qquad (3.28)$$

If s divides w, the Chinese Remainder Theorem guarantees that $\theta_{[s]}(x) \equiv \theta_{[w]}(x)$ modulo $x^s - 1$, and so

$$\theta_{[w]}(x) \frac{x^w - 1}{x^s - 1} \equiv \theta_{[s]}(x) \frac{x^w - 1}{x^s - 1} \qquad (\text{mod } x^w - 1)$$

thus (3.28) becomes

$$w\theta_{[w]}(x) \equiv w\theta_w(x) - \sum_{\substack{s|w \\ s \neq w}} \mu\left(\frac{w}{s}\right) s[\theta_{[s]}(x) - \theta_w(x)] \frac{x^w - 1}{x^s - 1} \quad (\text{mod } x^w - 1). \qquad (3.29)$$

Let $\pi = p_2 \dots p_j$, where the prime power decomposition of w is given by

$$w = p^a p_2^{a_2} \dots p_j^{a_j}$$

and define ω by $\omega = w/p\pi$. Then, since $\mu(q^2 r) = 0$ for any prime q, the only non-zero terms in the sum (3.29) are those for which s is divisible by ω. Hence dividing through by ω yields

$$p\pi\theta_{[w]}(x) \equiv p\pi\theta_w(x) - \sum_{\substack{r|p\pi \\ r \neq p\pi}} \mu\left(\frac{p\pi}{r}\right) r[\theta_{[r\omega]}(x) - \theta_w(x)] \frac{x^w - 1}{x^{r\omega} - 1} \qquad (3.30)$$

modulo $x^w - 1$. Thus the theorem is proved provided it can be shown that the sum on the right side of congruence 3.30 has all its coefficients divisible by $p\pi$.

By the second part of equation (3.23), with $w_1 = w/p^a$,

$$f_{p_2 \cdots p_j}(x^{\omega}) = f_{w_1}(x^{p^{a-1}}) \equiv f_{w_1}^{p^{a-1}}(x) \qquad \text{modulo } p.$$

Thus congruence 3.27 implies

$$\theta_w(x) - \theta_{[w/p]}(x) \equiv 0 \qquad \text{mod } [p, f_\pi(x^{\omega})]$$

which in turn implies, by congruence 3.24, that

$$[\theta_{[w/p]}(x) - \theta_w(x)]\, B_{p\pi,\pi}(x^{\omega}) \equiv 0 \qquad \text{mod}(p, x^w - 1)$$

or

$$[\theta_{[w/p]}(x) - \theta_w(x)] \sum_{r|\pi} \mu\left(\frac{\pi}{r}\right) r \frac{x^w - 1}{x^{r\omega} - 1} \equiv 0 \qquad \text{mod}(p, x^w - 1). \qquad (3.31)$$

As $\theta_{[r\omega]}(x) \equiv \theta_{[w/p]}(x)$ modulo $x^{r\omega} - 1$ for all divisors r of π, this becomes

$$\sum_{r|\pi} \mu\left(\frac{\pi}{r}\right) r[\theta_{[r\omega]}(x) - \theta_w(x)] \frac{x^w - 1}{x^{r\omega} - 1} \equiv 0 \qquad \text{mod}(p, x^w - 1)$$

where it should be noted that the polynomial on the left side of this congruence has integral coefficients by assumption. So, the unique integral polynomial $\Psi(x)$, of degree less than w, defined by

$$\sum_{\substack{r|p\pi \\ r \neq p\pi}} \mu\left(\frac{p\pi}{r}\right) r[\theta_{[r\omega]}(x) - \theta_w(x)] \frac{x^w - 1}{x^{r\omega} - 1} \qquad (\text{mod } x^w - 1)$$

has every coefficient divisible by $p\pi$. Thus

$$\theta_{[w]}(x) = \theta_w(x) - (p\pi)^{-1}\,\Psi(x)$$

has integral coefficients, as was to be shown.

Thus, provided a complete list of candidates for $\theta_d(x)$'s is available (for each divisor d of w) congruence 3.21 (or its computationally more convenient formulation, congruence 3.29) can be used together with Lemma 3.9 to construct integral polynomials $\theta_{[w]}(x)$ and ultimately, if these have non-negative coefficients, the difference set polynomial $\theta(x)$ itself.

Of course, for any divisor $d \neq 1$ of v, the major source of polynomials $\theta_d(x)$ is the equation

$$\theta_d(\zeta)\theta_d(\zeta^{-1}) = n \tag{3.32}$$

which must hold for any primitive d^{th} root of unity ζ. Thus there is a correspondence between the polynomials $\theta_d(x)$ and a restricted set of <u>principal</u> ideal factorizations of n over the field of dth roots of unity. If $c = \Sigma\, a_i\zeta^i$ is an algebraic integer which satisfies $c\bar{c} = n$ (the bar denotes complex conjugation) then, by a theorem of Kronecker [Theorem 2.20 above], the only $\theta_d(x)$'s associated with the principal ideal factorization $(c)(\bar{c}) = (n)$ are given by

$$\theta_d(x) \equiv \pm\, x^j \sum a_i x^i \qquad \text{modulo } f_d(x)\,. \tag{3.33}$$

Determining <u>all</u> such principal ideal factorizations of n in the field of dth roots of unity is, in general, an extremely difficult problem. However, for small parameters v, k, λ it can be done.

As an example of this method, the $v, k, \lambda = 21, 5, 1$ difference set $\{3, 6, 7, 12, 14\}$ will be constructed. Here $\theta_{[1]}(x) = k = 5$ and since (2) is a prime ideal in the field of 3rd roots of unity [Theorem 2.19 above], $\theta_3(x) = \pm\, 2x^a$

for some $a = 0,1,2$. By congruence 3.27

$$\theta_{[1]}(x) = 5 \equiv \pm 2x^a \equiv \pm 2 \qquad \text{mod}(3,\ x-1)$$

thus $\theta_3(x) = + 2x^a$ necessarily. By congruence 3.29

$$\theta_{[3]}(x) \equiv 2x^a + \frac{1}{3}[5 - 2x^a] \frac{x^3-1}{x-1} \equiv 2x^a + 1 + x + x^2 \qquad (\text{mod } x^3 - 1).$$

So, by shifting the difference set if necessary, $\theta_{[3]}(x) = 3 + x + x^2$ with the shift fixed modulo 3.

In the field of 7th roots of unity the ideal (2) splits into a product of two prime ideals [Theorem 2.19 above] and since, as Reuschle (1875, p. 7) lists, $(1 + \zeta_7^4 + \zeta_7^6)(1 + \zeta_7 + \zeta_7^3) = 2$, these ideals are principal. Thus, by equation (3.32), the ideal $(\theta_7(\zeta_7))$ can only be (2) or $(1 + \zeta_7^4 + \zeta_7^6)^2$ or $(1 + \zeta_7 + \zeta_7^3)^2$. Since these last two could only correspond to equivalent difference sets, only one of them need be considered. If $(\theta_7(\zeta_7)) = (2)$ then $\theta_7(x) = \pm 2x^b$ $(b = 0,1,\ldots,6)$ and congruence 3.27 shows that the sign is negative. However using $\theta_7(x) = -2x^b$ in congruence 3.29 yields negative coefficients for $\theta_{[7]}(x)$, a contradiction. Thus, without loss of generality, one may assume that $\theta_7(x) = \pm (1 + x^4 + x^6)^2 x^c$ for some $c = 0,1,\ldots,6$. By congruence 3.27

$$5 \equiv \pm (1 + x^4 + x^6)^2 x^c \equiv \pm 9 \qquad \text{mod}(7,\ x-1)$$

thus the negative sign prevails and from congruence 3.29

$$\theta_{[7]}(x) \equiv -x^c(1 + x + 2x^3 + 2x^4 + x^5 + 2x^6) + 2(1 + x + \cdots + x^6) \qquad (\text{mod } x^7 - 1).$$

The different values of c correspond to different shifts of the set. Since $(3,7) = 1$, c can be specified arbitrarily without affecting $\theta_{[3]}(x)$; thus by uniquely specifying the shift one may assume that $(c = 5)$

$$\theta_{[7]}(x) = 2 + x^3 + x^5 + x^6 \quad \text{and} \quad \theta_{[3]}(x) = 3 + x + x^2 .$$

In the field of 21st roots of unity, the ideal (2) splits into a product of two prime ideals [Theorem 2.19 above] and since $(1 + \zeta_{21}^{12} + \zeta_{21}^{18})(1 + \zeta_{21}^{3} + \zeta_{21}^{9}) = 2$, these ideals are principal. As before, the ideal $(\theta_{21}(\zeta_{21}))$ can only be (2) or $(1 + \zeta_{21}^{3} + \zeta_{21}^{9})^2$ or $(1 + \zeta_{21}^{12} + \zeta_{21}^{18})^2$. The first two of these do not satisfy congruences 3.27; thus $\theta_{21}(x) = \pm (1 + x^{12} + x^{18})^2 x^d$ for some $d = 0,1,\ldots,20$. When $p = 7$ congruence 3.27 rules out the minus sign and shows that $d = 3e$. For $p = 3$, this same congruence yields $e = 5$. Thus $\theta_{21}(x) = (1 + x^{12} + x^{18})^2 x^{15} \equiv x^9 + x^{15} + x^{18} + 2(x^3 + x^6 + x^{12})$ modulo $x^{21} - 1$. Computing $\theta_{[21]}(x)$ by congruence 3.29 yields

$$\theta(x) = \theta_{[21]}(x) \equiv x^3 + x^6 + x^7 + x^{12} + x^{14} \qquad (\text{mod } x^{21} - 1)$$

i.e., $\{3, 6, 7, 12, 14\}$ is the desired difference set.

Of course, there are much simpler ways to construct this particular difference set (Lemma 3.7 for example), but the method is perfectly general and works whenever all the principal ideal factorizations of (n) in the fields of dth roots of unity can be determined. Thus there are parameter sets where this method is easier to apply than any of the others. As Turyn (1960) noted, the cases where n is a square and $\theta_d(x) = \pm \sqrt{n}\, x^s$ for all divisors d of v are particularly nice.

Note 1. In the course of the example above the fact that a trial $\theta_{[7]}(x)$ had a negative, though integral, coefficient was used to exclude the possibility $\theta_7(x) = -2x^b$. This is a consideration outside the range of Lemma 3.9. Thus one might be led to suspect that the complete collection of conditions (3.27) imposed by Lemma 3.9 was not sufficient to guarantee the existence of a difference set. This is false. That is, given that the polynomials $\theta_d(x)$ used are meaningful for the problem, i.e., that

$$\theta_d(x)\theta_d(x^{-1}) = \begin{cases} n & \text{when } x = \zeta_d \quad d \neq 1 \\ k^2 & \text{when } x = 1 \end{cases} \tag{3.34}$$

then any integral polynomial $\theta_{[v]}(x)$ computed by the above process is the Hall polynomial of a v, k, λ - difference set. For, whatever else it is, it is an integral polynomial of degree less than v which satisfies

$$\theta_{[v]}(x)\theta_{[v]}(x^{-1}) \equiv n + \lambda(1 + x + \cdots + x^{v-1}) \qquad (\text{mod } x^v - 1)$$

$$\theta_{[v]}(1) = k$$

as follows from equations (3.26) and (3.34). Thus with $\theta_{[v]}(x) = \Sigma\, a_i x^i$ (a_i integers) it follows that (compare constant coefficients)

$$a_0^2 + a_1^2 + \cdots + a_{v-1}^2 = k$$

$$a_0 + a_1 + \cdots + a_{v-1} = k$$

and the only solutions to these diophantine equations have $a_i = 0,1$. So the conditions (3.27) of Lemma 3.9 together with the trivially necessary conditions (3.34) are not only necessary but also sufficient for the existence of a v, k, λ - difference set.

Note 2. Since this method and that based on Lemma 3.8 are both aimed either at establishing the non-existence or at the construction of successive polynomials $\theta_{[w]}(x)$, they are often combined (i.e., for a particular value of w whichever method is easier takes precedence). Generally speaking however (as seen in the example given in section III.C.) the successful application of Lemma 3.8 requires the knowledge of a multiplier or at least a w-multiplier. [Thus, in view of the multiplier theorems of section III.A, Lemma 3.8 is more likely to be useful when

v has a relatively large divisor w prime to n.] On the other hand, multipliers play no obvious role in the method of this section [where known they can be used to restrict the possibilities for $\theta_d(x)$]; thus the two approaches tend to complement each other.

Note 3. The successive construction of the $\theta_{[w]}(x)$'s as a means of determining whether or not a particular difference set might exist has been used almost from the beginning of the study of difference sets. Indeed, the use of the algebraic number theoretic implications of congruence 3.32 in this, is also quite standard. [Perhaps the best documented examples are in the works of Turyn (1960, 1961).] Nevertheless the explicit determination of the relations (3.19), (3.20), (3.21) upon which the method is based as well as Lemma 3.9 is quite recent and is due to H. C. Rumsey, Jr.

IV. DIFFERENCE SETS OF SPECIAL TYPE

Various groupings of difference sets have been studied more extensively than others. These groupings usually consist of all (or all cyclic) difference sets with a certain fixed property. For example, those having $\lambda = 1$ or those which may be constructed by some special process. If the common property is of a constructive nature, the grouping is usually called a family of difference sets. (These difference set families and their special constructions are discussed in Chapter V.) The present chapter concerns itself with some groupings of difference sets which have received special attention but are not of the familial type.

A. Planar Difference Sets

The incidence matrix of a non-trivial symmetric block design with $\lambda = 1$ is also the incidence matrix of a finite projective plane. That is, if the blocks are called lines and the objects are called points, the incidence matrix details the structure of a system of $v = n^2 + n + 1$ points and v lines such that

(i) each line contains exactly $n + 1$ points and each point is on exactly $n + 1$ lines

(ii) any two distinct points are contained in one and only one line; any two distinct lines contain one and only one point in common

(iii) there exist four points no three of which are on the same line.

This last condition serves to exclude certain trivial configurations. [An introduction to the study of finite projective planes is provided by Albert and Sandler (1968). See Dembowski (1968) for a comprehensive survey.]

An open question, which has received a great deal of attention is that of determining the values n for which finite projective planes, with $v = n^2 + n + 1$ points, exist. They are known to exist whenever n is a prime power and known not to exist whenever the associated Bruck-Ryser condition (see section II.B. for this) is not satisfied. For all other values of n, the existence of a finite projective plane is undecided. [In particular $n = 10$ is undecided; i.e., it is not known whether or not a symmetric block design exists with parameters $v, k, \lambda = 111, 11, 1$.]

Primarily because of the interest in this problem, the existence question for cyclic symmetric block designs with $\lambda = 1$ (i.e., difference sets with $\lambda = 1$) has been pursued extensively. These difference sets are called planar or simple. Planar difference sets do exist with parameters $v = p^{2j} + p^j + 1$, $k = p^j + 1$, $\lambda = 1$ for all prime powers $p^j = n$ [Singer (1938), see section V.A. for construction details]. On the other hand, not all finite projective planes correspond to cyclic difference sets; those that do are called finite cyclic projective planes.

Many of the results originally developed for planar difference sets were subsequently generalized to the case of arbitrary λ and as such appear in earlier sections of this survey. In order to increase the readability of this section some of these results are repeated here; others are merely referred to when needed.

Three areas of interest regarding these planar difference sets are discussed in this report. Singer's construction process for planar difference sets with $n = p^j$, p prime, is presented in section V.A. The question (still open) of whether there can exist multiple inequivalent planar difference sets for prime power n is mentioned in section VI.A and elsewhere. Finally, in this section, results concerning planar difference sets for general values of n (i.e., not restricted to prime powers) are given. Of course, since all known planar difference sets are of the Singer type, these results are mainly rules which establish the non-existence of planar difference sets with certain parameter values.

If t is a multiplier of a planar difference set, then t determines an automorphism of the associated symmetric block design (as noted in section I.E. above) and hence an automorphism of the associated finite cyclic projective plane π; thus t is also said to be a multiplier of the plane π. [If the ordering of the points of π under its cyclic automorphism is $P_0, P_1, \ldots, P_{v-1}$, then, in order for an arbitrary automorphism α to be a multiplier of π, it is necessary that there exists an integer t such that the points of the plane are permuted under α according to the rule $P_i \to P_{ti}$.]

All non-trivial planar difference sets have non-trivial multipliers as is easily seen from Theorem 3.1 and the parameters $v, k, \lambda = n^2 + n + 1, n + 1, 1$. In particular, all divisors t of n are multipliers. [In fact, the primes 2 (Hall, 1947) and 3 (Mann, 1952) are multipliers of a planar difference set if and only if they divide n. This is not true in general; for 11 ($\equiv 2^5$) is a multiplier of the Singer set $v, k, \lambda = 21, 5, 1$.] Since $(k,v) = 1$ for all planar difference sets, Theorem 3.6(2) shows that there exists at least one shift E of the difference set (i.e., line of the plane) which is fixed by all the multipliers. Actually, combining a result of Evans and Mann (1951) with one of Mann (see Hall, 1947):

Theorem 4.1. At least one and at most three shifts of any planar difference set are fixed by all the multipliers. When $n \equiv 0,2$ modulo 3 there is a unique fixed shift E; this shift contains the object 0 for $n \equiv 0$ modulo 3 and does not contain 0 when $n \equiv 2$ modulo 3. If $n \equiv 1$ modulo 3 then $v = 3v_1$ and there may be one or three shifts fixed by all the multipliers. The shift determined by the pair of objects $v_1, 2v_1$ is always fixed, whereas the shifts determined by the object pairs $0, v_1$ and $0, 2v_1$ are fixed if and only if every multiplier satisfies $t \equiv 1$ modulo 3.

Proof. Since $v = (n - 1)(n + 2) + 3$ it follows that $(n - 1, v) = 1$ or 3. Thus applying Theorems 3.4 and 3.5 to the multiplier n yields the first part of the theorem. When $n \equiv 0, 2$ modulo 3 then $(n - 1, v) = 1$; thus the fixed shift is unique and no divisor of v also divides $n - 1$. This implies that all the blocks (save {0}) fixed by the multiplier n contain 3 elements each, e.g., $\{a, na, n^2a\}$. Thus determining whether or not 0 belongs to the unique fixed shift E. When $n \equiv 1$ modulo 3 then $(n - 1, v) = 3$; so $v = 3v_1$. Here n, as well as any other multiplier $t \equiv 1$ modulo 3, fixes the three objects $0, v_1, 2v_1$ and hence also the three shifts containing the object pairs $0, v_1$; $0, 2v_1$; $v_1, 2v_1$. Since $(t,v) = 1$ necessarily, the only other multipliers possible have $t \equiv 2$ modulo 3. These fix 0 and interchange v_1 with $2v_1$; thus the only shift fixed by n and also fixed by such a multiplier is the one containing v_1, $2v_1$. Thus the theorem is proved.

With $n = 2,3,4,7$ the planar difference sets of Singer (see section V.A.) provide an example for each of the possibilities listed in Theorem 4.1.

Corollary 4.2 (Evans and Mann, 1951). If t is a multiplier of a non-trivial planar difference set with $v = n^2 + n + 1$ prime, then the order α of t modulo v divides one of n, $n + 1$. If $n \equiv 0$ modulo 3 then α divides n while if $n \equiv 2$ modulo 3 then α divides $n + 1$.

Proof. Since v is prime it follows that $n \not\equiv 1$ modulo 3 (for in this case 3 divides v). So $n \equiv 0$, 2 modulo 3 and hence there is a unique shift E fixed by all the multipliers. Furthermore (since v is prime), all the non-zero residues of v are distributed into disjoint blocks $\{a, ta, \ldots, t^{\alpha-1}a\}$ of uniform size α. As E is necessarily a union of these blocks with the possible addition of the object 0 the only question is whether or not 0 belongs to E. So the corollary follows from Theorem 4.1.

Another important result, used but not explicitly stated by Hall (1947), was stated by Mann (1952) as

Theorem 4.3. If t_1, t_2, t_3, t_4 are multipliers of a planar difference set and if $t_1 - t_2 \equiv t_3 - t_4$ modulo v, then

$$(t_1 - t_2)(t_1 - t_3) \equiv 0 \qquad \text{modulo } v\,.$$

Proof. Theorem 4.1 guarantees the existence of a shift E fixed by all multipliers. Such a shift must contain, together with each of its elements e, the elements t_1e, t_2e, t_3e, t_4e; thus

$$t_1e - t_2e \equiv t_3e - t_4e \qquad \text{modulo } v\,.$$

Since $\lambda = 1$ this is only possible if $t_1e = t_2e$ or $t_1e = t_3e$. So for every element e of E it follows that

$$(t_1 - t_2)(t_1 - t_3)e \equiv 0 \qquad \text{modulo } v. \qquad (4.1)$$

Now $1 \equiv e_i - e_j$ modulo v for some e_i, e_j of E; thus subtracting the associated equations (4.1) for e_i and e_j establishes the theorem.

Applying Theorem (4.3) with $t_1 = 1$, $t_2 = 2$, $t_3 = 2^j$, $t_4 = 2^j + 1$ shows that the only planar difference sets having 2 and $2^j + 1$ as multipliers are those for which they are not distinct multipliers, i.e., $2^j - 1 \equiv 0$ modulo v. Similarly $t_1 = 1$, $t_2 = 2$, $t_3 = 2^j - 1$, $t_4 = 2^j$ yields the result that the only planar difference sets having 2 and $2^j - 1$ as multipliers are those for which $2^j - 2 \equiv 0$ modulo v. When Theorem 4.3 is combined with the fact that every divisor t of n is a multiplier for any planar difference set (Theorem 3.1), it provides a quite effective non-existence test for these difference sets in the case where n is composite. Many of these cases fall within the scope of the following corollary:

Corollary 4.4. (Evans and Mann, 1951) Let a_1, a_2, a_3 be non-negative integers for which

$$q - p_1^{a_1} = p_2^{a_2} - p_3^{a_3}, \quad p_1^{a_1} < 3q, \quad p_2^{a_2} < 3q$$

where q, p_1, p_2, p_3 are prime divisors of n and $p_1 \neq q \neq p_2$. Then there are no planar difference sets with $v = n^2 + n + 1$.

Proof. Since $n \geq p_1q \geq 2q$ it follows that the hypotheses imply

$$0 < \left| \left(q - p_1^{a_1}\right) \left(q - p_2^{a_2}\right) \right| < 4q^2 \leq n^2 < v$$

which contradicts Theorem 4.3 and establishes the corollary.

This corollary shows, for example, that no planar difference set has $n \equiv 0$ modulo 6 [let $q = 3$, $p_1 = p_2 = p_3 = 2$, $a_1 = a_2 = 1$, $a_3 = 0$].

Another efficient non-existence test for planar difference sets having composite n is provided by the next theorem. [Actually it is no more than a collection of some of the more easily recognizable special cases of Theorem 2.9. Of course, these special cases were established before Theorem 2.9 was known in full generality. [See Hall (1947), Mann (1952), Hall and Ryser (1951) and Evans and Mann (1951)].

Theorem 4.5. Let t be a multiplier of a planar difference set and let p and q be prime divisors of n and v respectively. Then if any of the following conditions are met n is necessarily a square:

(i) t has even order modulo q

(ii) p is a quadratic non-residue of q

(iii) $n \equiv 4$ or 6 modulo 8

(iv) $n \equiv 1$ or 2 modulo 4 and $p \equiv 3$ modulo 4

(v) $n \equiv n_1$ or n_1^2 modulo $n_1^2 + n_1 + 1$ and p is of even order modulo $n_1^2 + n_1 + 1$.

Proof. The elementary number theory of quadratic residues is used in this proof [see, for example, Nagell (1951)].

(i) If the order of t is $2f$ then $t^f \equiv -1$ modulo q, so Theorem 2.9 and the fact that $(n,v) = 1$ suffice.

(ii) Every non-residue of q is of even order modulo q.

(iii) If $n \equiv 4,6$ modulo 8 then n is even (let $p = 2$) and $v \equiv 5,3$ modulo 8. So the Jacobi symbol $\left(\frac{2}{v}\right) = -1$. Thus there exists at least one prime divisor q of v for which 2 is a quadratic non-residue, i.e., this reduces to (ii).

(iv) By the reciprocity law for the Jacobi symbol and the hypotheses above

$$\left(\frac{p}{v}\right) = -\left(\frac{v}{p}\right) = -\left(\frac{n^2+n+1}{p}\right) = -1$$

since p divides n. This insures the existence of a proper prime divisor q of v as in (iii).

(v) Since $n \equiv n_1$ or n_1^2 it follows that $v \equiv 0$ modulo $n_1^2 + n_1 + 1$. So any divisor of $n_1^2 + n_1 + 1$ is also a divisor of v. Since p is of even order modulo $n_1^2 + n_1 + 1$ it is necessarily of even order modulo some prime divisor q of $n_1^2 + n_1 + 1$; thus this case reduces to (i) and the theorem is proved.

As an example consider case (v) of Theorem 4.5 when $n_1 = 3$. This shows that n must be a square if a planar difference set is to exist with $n \equiv 3$ or 9 modulo 13 and $p \equiv 2,4,5,6,7,8,10,11$ or 12 modulo 13. Among others this excludes the possibility $n = 35$.

Whereas Theorem 4.1 shows that at most 3 shifts of any planar difference set are fixed by <u>all</u> the multipliers, an individual multiplier t may fix more than 3 shifts. In fact

<u>Theorem 4.6.</u> (Hall, 1957). Let t be a multiplier of a planar difference set D and let π designate the finite cyclic projective plane generated by D. Then there are exactly $(t - 1, v) = v_1$ objects and exactly v_1 shifts of D fixed by the multiplier t. Of course v_1 may be 1 or 3 but if $v_1 > 3$ then $v_1 = n_1^2 + n_1 + 1$ and the fixed objects together with the fixed shifts determine a cyclic subplane π_1 having $n_1 + 1$ points on a line. Further, every multiplier of D is a multiplier of the subplane π_1.

<u>Proof.</u> The first assertion was established for general λ in section III.B. above. Assume $v_1 > 1$ let $v = v_1 v'$, then the fixed elements are $0, v', \ldots, (v_1 - 1)v'$. Let $E = \{e_1, \ldots, e_{n+1}\}$ be a shift of D which is fixed by the multiplier t. Then E is of course also a planar difference set. So the representation $\ell v' = e_i - e_j$ is unique $(1 \leq \ell < v_1)$. But since t fixes E, as well as all multiples of v', it follows that $e_i - e_j = te_i - te_j$ are distinct representations from E of $\ell v'$ unless t also fixes e_i and e_j. Hence every non-zero multiple of v' is uniquely represented as a difference of elements of E; these elements from E being multiples of v' also. So the elements of E

divisible by v' determine a planar difference set (non-trivial for $v_1 > 3$) with parameters $v_1 = n_1^2 + n_1 + 1$, $k_1 = n_1 + 1$. Since each shift fixed by t is of the form $E + jv'$ for some j, the objects fixed by t together with the shifts fixed by t constitute a cyclic subplane π_1 with $n_1 + 1$ points on each line as was to be proved. If s is another multiplier of D then s permutes the shifts fixed by t among themselves (Theorem 3.5) hence s is a multiplier of the subplane π_1 also. Thus all assertions of the theorem have been established.

Bruck (1955) has shown that if a finite projective plane π, with $v = n^2 + n + 1$, has a proper subplane π_1, with $v_1 = n_1^2 + n_1 + 1$, then $n = n_1^2$ or $n \geq n_1^2 + n_1$. [For, each line of π_1 contains $n - n_1$ points of π which are not points of π_1. Thus the number of points of π which lie on no line of π_1 is

$$n^2 + n + 1 - (n_1^2 + n_1 + 1) - (n - n_1)(n_1^2 + n_1 + 1) = (n - n_1)(n - n_1^2) .$$

Since $n > n_1$ necessarily, it follows that $n \geq n_1^2$. If n is actually greater than n_1^2 there exists a point P of π which lies on no line of π_1. So the $n_1^2 + n_1 + 1$ lines joining P to the points of π_1 are distinct and as P lies on exactly $n + 1$ lines of π it must be that $n \geq n_1^2 + n_1$.] Roth (1964) shows that in certain special cases (including that of Theorem 4.6) this inequality may be improved to $n \geq n_1^2 + n_1 + 2$. In all known cases of Theorem 4.6 n_1 divides n, but the question of whether this must be so is open. (For an arbitrary subplane of an arbitrary finite projective plane examples are known where n_1 does not divide n, see H. Neumann, 1955).

For any integer $j \geq 0$, t^j is a multiplier whenever t is, and t^j certainly fixes every object and shift that t fixes. If E is any shift fixed by t and if $(t^j - 1, v) = v_0$ then Theorem 4.6 shows that $v_0 = n_0^2 + n_0 + 1$ and that there are thus $n_0 + 1$ objects of E fixed by t^j. So

<u>Corollary 4.7</u> (Hall, 1947). If t is a multiplier of a cyclic plane π, if E is a line fixed by t and if v_0 is a divisor of v such that $(t^x - 1, v) = v_0$

for $x = j$ but not for $x < j$, then $v_0 = n_0^2 + n_0 + 1$ and there are exactly $n_0 + 1$ points on E fixed by t^j. These $n_0 + 1$ points of E are thus permuted by t in cycles whose lengths divide j.

Corollary 4.8 (Ostrom, 1953). If a planar difference set D has parameters $v = m^{2r} + m^r + 1$, $n = m^r$, with $(r,3) = 1$ then there exists a planar difference set D_1 with parameters $v_1 = m^2 + m + 1$, $n_1 = m$. Furthermore, every multiplier of D is also a multiplier of D_1.

Proof. Since m^3 is obviously a multiplier of D, the result follows from Theorem 4.6 once it can be established that

$$(m^3 - 1, m^{2r} + m^r + 1) = m^2 + m + 1 = v_1 . \tag{4.2}$$

From $(r,3) = 1$ it follows that v_1 divides the left side of equation (4.2), thus the only question is whether some multiple of v_1 does also. Now $v = (n-1)\cdot(n+2) + 3$ shows that $(n - 1, v) = 1$ or 3; thus $(m - 1, v) = 1$ or 3. So $3v_1$ is the only candidate. If it were $3v_1$ then $m \equiv 1$ modulo 3 necessarily and this implies that $v_1 \equiv 0$ modulo 3. Thus $v \equiv 0$ modulo 9 and there is no value of n for which this happens. So equation (4.2) is valid and the corollary has been established.

Assume a planar difference set D exists and let q be a prime divisor of v. Then q is said to be a type I divisor of v if there exists some multiplier t of D having lower order modulo q than it does modulo v. Let this order modulo q be α. Then

$$(t^{\alpha} - 1, v) = w \equiv 0 \qquad \text{modulo } q$$

where w properly divides v. By Theorem 4.6 $w = n_1^2 + n_1 + 1$ and either $w = 3$ or w is the modulus of a proper subplane π_1. Thus, in particular (Evans and Mann, 1951):

Corollary 4.9. If no proper factor of v divisible by the prime q is the modulus of a planar difference set, then the order of every multiplier t modulo q is also its order modulo v.

Prime divisors q and v for which every multiplier t has the same order as it does for v are called type II divisors.

Before going further it should be noted that the only prime divisors q of $n^2 + n + 1$ are $q = 3$ and $q \equiv 1$ modulo 3. For if $n^2 + n + 1 \equiv 0$ modulo q then $n^3 - 1 \equiv 0$ modulo q and this implies that the order α of n modulo q divides 3, i.e., $\alpha = 1$ or 3. If $\alpha = 1$ then $n \equiv 1$ modulo q and $n^2 + n + 1 \equiv 3$ modulo q; so $q = 3$. If $\alpha = 3$ then 3 divides $\varphi(q)$ i.e., $q \equiv 1$ modulo 3.

Theorem 4.10 (Ostrom, 1953). If a planar difference set D has a type II prime divisor q of v, then the multiplier group of D is cyclic and its order divides $q - 1$.

Proof. The image under reduction modulo q of the multiplier group of D is of course a subgroup of the non-zero residues of q. Hence this image group is cyclic. No two multipliers have the same image. For let t_1, t_2 be distinct multipliers of D (i.e., $t_1 \not\equiv t_2$ modulo v). Then there exists a multiplier t_3 such that $t_2 \equiv t_1 t_3$ modulo v. Thus $t_2 \equiv t_1 t_3$ modulo q and if then $t_1 \equiv t_2$ modulo q, it would follow that $t_1 (t_3 - 1) \equiv 0$ modulo q. Since $(t_1, q) = 1$ necessarily this implies that $t_3 - 1 \equiv 0$ modulo q. But q is of type II, so $t_3 - 1 \equiv 0$ modulo v also and this contradicts the distinctness of t_1, t_2. So the multiplier group of D is isomorphic with a subgroup of the non-zero residues of q i.e., it is cyclic and has order dividing $q - 1$.

Corollary 4.8 shows that if a planar difference set with $n \neq p^j$ is sought there is no loss of generality in restricting the search to the cases where n is not a square. When this is done, Theorem 4.5 insures that no multiplier is of even order modulo any of the prime divisors q of v. Thus

Theorem 4.11 (Ostrom, 1953). Suppose that there exists a planar difference set modulo $v = n^2 + n + 1$ where $v = q_1 q_2 \ldots q_j$ and the q_i are not necessarily distinct primes. Suppose further that n is not a square and that for some i, q_i is of type II. Then the order S of the cyclic group of multipliers is $3h$ where h is odd and $3h$ divides $\varphi(q_i) = q_i - 1$.

Theorem 4.12 (Ostrom, 1953). If the hypotheses of Theorem 4.11 are satisfied and all the prime divisors q_i of v are of type II, then the order S of the cyclic group of multipliers divides $n + 1$ when $n \equiv 2$ modulo 3 and S divides n when $n \equiv 0$ modulo 3. (When $n \equiv 1$ modulo 3, then 3 is always a type I divisor of v).

Proof. By Theorem 4.10 the multiplier group is cyclic of order S; let t be a generator of this group. Since every prime divisor of v is of type II the non-zero residues of v are distributed by t into disjoint sets $\{a, ta, \ldots, t^{S-1}a\}$ of uniform size S. Thus the unique (Theorem 4.1) shift fixed by all the multipliers is necessarily a union of these sets together with the possible addition of the object 0. By Theorem 4.1 the object 0 is added if and only if $n \equiv 0$ modulo 3. Thus the theorem is proved.

Combining a result of Ostrom (1953) with one of Evans and Mann (1951) yields

Theorem 4.13. Suppose that there exists a planar difference set with $v = v_1 v_2$, $v_1 > 1$ and $v_2 > 1$, where every prime divisor of v_2 is of type II with respect to v. Suppose further that $(t^j - 1, v) = v_1$, where $j < S$ and t is a generator of the cyclic group of multipliers. Then $v_1 = n_1^2 + n_1 + 1$ and S divides $n - n_1$. If in addition v_1 and v_2 are primes, $n_1 \neq 1$ and v_1 divides $n - n_1$ then S divides $(n - n_1)/v_1$.

Proof. Let E be a shift fixed by t^j; it contains $n_1 + 1$ objects which are multiples of v_2 (see the proof of Theorem 4.6 above). If a is any of the other $n - n_1$ objects of E, let α be the least power of t such that $a^t \equiv a$ modulo v. Then $a(t^\alpha - 1) \equiv 0$ modulo $v_1 v_2$ hence $t^\alpha - 1 \equiv 0$ modulo some type II divisor q of v_2. Thus $\alpha = S$ and S divides $n - n_1$. If v_2 is prime then (Theorem 4.10) S divides $v_2 - 1$. Since $n^2 + n + 1 = (n_1^2 + n_1 + 1)v_2$

$$v_2 - 1 = \frac{(n - n_1)(n + n_1 + 1)}{v_1} .$$

Now v_1 divides $n - n_1$ by assumption, thus $n + n_1 + 1 \equiv 2n_1 + 1 \not\equiv 0$ modulo v_1 (since $n_1 \neq 1$). Since S divides $n - n_1$ and v_1 is prime it follows that S divides $(n - n_1)/v_1$ as was to be shown.

Utilizing mainly Theorem 4.3 Hall (1947) established that the only planar difference sets with $n \leq 100$ have n a prime power. This was extended by Evans and Mann (1951) to $n \leq 1600$ and according to Dembowski (1968, p. 209) was pushed to $n \leq 3600$ by V. H. Keiser (unpublished).

The tests of Evans and Mann. Let t be a multiplier and let p, q be prime divisors of n, v respectively. Then the existence of a difference set with parameters $v, k, \lambda = n^2 + n + 1, n + 1, 1$ is subject to the following conditions:

(a) Corollary 4.4 and more generally Theorem 4.3 must hold.

(b) If $n \equiv 4,6$ modulo 8 then n must be a square.

(c) If $n \equiv 1,2$ modulo 4 and if $p \equiv 3$ modulo 4 then n is a square.

(d) If $n \equiv n_1$ or n_1^2 modulo $n_1^2 + n_1 + 1$ and if p is of even order with respect to $n_1^2 + n_1 + 1$, then n is a square. (In particular the associated tests for $n_1 = 1,2,3,5,6,7$ were used.)

(e) If p is a quadratic non-residue of q then n is a square.

(f) If t is of order α modulo v and v is prime, then α divides n if $n \equiv 0$ modulo 3. If n is not a square α must be odd.

(g) If t is of order α modulo v and v is prime, then α divides $n + 1$ if $n \equiv 2$ modulo 3. If n is not a square α must be odd.

(h) Let $v = v_1 q$ where no planar difference set exists modulo v_1. Then the order α of any multiplier t modulo q must divide $\beta = (\varphi(v_1), \varphi(q))$. If n is not a square α must be odd.

(i) Let $v = v_1 q$, where v_1, q are both primes and a planar difference set exists modulo v_1 but not modulo q. Let there exists a non-trivial multiplier t of the $v,n+1,1$ - difference set such that $t \equiv 1$ modulo v_1 but $t \not\equiv 1$ modulo q. If n is not a square then the order of t modulo q divides $n - n_1$ and is odd. If, in addition, v_1 divides $n - n_1$ $(n_1 \neq 1)$ then the order of t modulo q divides $(n - n_1)/v_1$ and is odd.

Proof. Theorem 4.3, Corollary 4.4 and Theorem 4.5 establish (a),...,(e) above. Tests (f) and (g) are a combination of Corollary 4.2 and Theorem 4.5. Consider test (h). Here Theorem 4.6 shows that the order of t modulo v_1 is the same as its order modulo v thus α divides $\varphi(v_1)$ and α divides $\varphi(q)$ by definition. So Theorem 4.5 finishes off. Consider test (i). Since q and v_1 are prime it follows from Corollary 4.9 that q is a type II divisor. The multiplier t of the test corresponds to the multiplier t^j of Theorem 4.13. Thus Theorems 4.5 and 4.13 establish the validity of test (i).

Since the only known planar difference sets have prime power n and in fact may all be constructed by the process of Singer (see Chapter V), it is easy to make up perfectly reasonable conjectures about planar difference sets. Just conjecture that any property possessed by the Singer sets holds for all planar difference sets. One such property of the Singer set $v,k,\lambda = p^{2j} + p^j + 1, p^j + 1, 1$ is that its multiplier group consists of all the powers of p modulo v. [That all powers of p are necessarily multipliers has been observed earlier. The fact that only the powers of p are multipliers is non-trivial and is due to Gordon, Mills and Welch (1962), see section V.A. for a proof of their more general result.] For a general planar difference set (i.e., n not necessarily a prime power) the analogous conjecture would be that its multiplier group was generated by the prime divisors of n. A reduction lemma established by Halberstam and Laxton (1964), in the course of providing an alternate proof ($\lambda = 1$ only) of the Gordon, Mills, Welch result on Singer sets, may be of some use in attacking this problem. Note that t, nt, n^2t are necessarily either all multipliers or all non-multipliers of any planar difference set with $k = n + 1$. Furthermore, define an integer j to be of reduced type modulo $n^2 + n + 1$ if

$$j \equiv r + sn \qquad \text{modulo } n^2 + n + 1$$

with $0 \leq r, s < n$ and $0 < r + s \leq n$. Then

Lemma 4.14. (Halberstam and Laxton) If $t > 1$ and if $(t, n^2 + n + 1) = 1$ with $n > 1$, then at least one of t, nt, n^2t is of reduced type modulo $n^2 + n + 1$.

B. Hadamard Difference Sets

Difference sets whose parameters $v, k, \lambda = 4t - 1, 2t - 1, t - 1$ are called Hadamard difference sets. Like the planar difference sets of section IV.A., these have been extensively studied. There are several reasons for this; among them are:

(i) the relative abundance of such difference sets

(ii) with $k < v/2$ as usual, λ varies between 1 and $(v - 3)/4$, see section I.B. Thus planar difference sets and Hadamard difference sets present the extreme values of λ.

(iii) the autocorrelation function of the 1, -1 characteristic function of these difference sets (see section I.D. for this) is minimal [$R_b(j) = -1$ for $j \not\equiv 0$ modulo v]. This has led to several digital communications applications [see, for example, Golomb et al (1964) or Goldstein (1964)].

(iv) the relation between these difference sets and the, as yet unsolved, Hadamard matrix problem (see the note below).

The known Hadamard difference sets can be classified according to the value of v. The groupings are:

(a) $v = 2^j - 1$, $j \geq 2$; section V.A. discusses a large family of difference sets whose parameters include these (construction details are given there).

(b) $v = 4t - 1$ is prime; here there always exists a Hadamard difference set composed of the quadratic residues modulo v (see section V.B.) and when the prime $v = 4t - 1$ is expressible as $4x^2 + 27$, there is an additional difference set (due to Hall, 1956) whose construction is discussed in section V.C. Some others exist also.

(c) $v = p(p + 2)$ where p and $p + 2$ are both prime numbers (see section V.D. for the details).

Occasionally v satisfies more than one of these conditions and, for the most part, this leads to multiple inequivalent difference sets. Specifically:

(a) and (b) overlap if and only if v is a Mersenne prime; the only Mersenne primes of the form $4x^2 + 27$ are $v = 31$, 127 and 131071 [see Skolem, Chowla and Lewis (1959) for this].

(a) and (c) overlap if and only if $v = 15$.

The known Mersenne primes (i.e., primes of the form $2^j - 1$) are $v = 2^j - 1$ with $j = 2, 3, 5, 7, 13, 17, 19, 31, 61, 89, 107, 127, 521, 607, 1279, 2203, 2281, 3217, 4253, 4423, 9689, 9941, 11213$ [see Gillies, 1964; in fact $2^{11213} - 1$ is, at present, the largest number known to be prime].

The difference sets corresponding to these v's are inequivalent except for $v = 3, 7, 15$ and 31. Now $v = 31 = 2^5 - 1 = 4.8 - 1 = 4.1^2 + 27$ is prime, thus (a) provides a difference set and (b) provides two more. However, the difference set corresponding to $4x^2 + 27$ is equivalent to the one from (a); thus only two truly distinct difference sets arise from $v = 31$.

While all known Hadamard difference sets have v's of types (a), (b), (c), not all of them can be constructed by the methods (section V.A. - V.D.) indicated in (a), (b), (c). In particular, there are exactly six inequivalent difference sets with parameters $v, k, \lambda = 127, 63, 31$ (see Baumert and Fredricksen, 1967), three of which do not arise from these constructions.

It is known (Golomb, Thoene, Baumert) that if a Hadamard difference set has $v < 1000$, then v is one of the forms (a), (b), (c) with six possible exceptions.

These exceptions are $v = 399, 495, 627, 651, 783$ and 975. The methods used to decide this were, of course, those of Chapter II and III.

Some problems of electrical network theory [Belevitch (1968), see also Goethals and Seidel (1967)] have led to the consideration, among other things, of skew Hadamard difference sets. That is, Hadamard difference sets which contain precisely one of the residues $d, v - d$ for $1 \le d \le v - 1$. Of course, the obvious examples of such sets are the quadratic residues (see section V.B.) of primes $q \equiv 3 \pmod 4$. Johnsen (1966B) has shown that there are no others.

Theorem 4.15. (Johnsen) The only cyclic difference sets which are skew Hadamard are given by the quadratic residues of a prime q, where $q \equiv 3 \pmod 4$.

Proof. Let $T(x) = 1 + x + \cdots + x^{v-1}$ and consider the Hall polynomial $\theta(x)$ for such a difference set. It satisfies $1 + \theta(x) + \theta(x^{-1}) \equiv T(x)$ modulo $x^v - 1$ and so [using $\theta(x)\theta(x^{-1}) \equiv n + \lambda\, T(x)$ and $\theta(x)T(x) \equiv k\, T(x)$] it follows that

$$\theta^2(x) + \theta(x) + n \equiv n\, T(x) \qquad \mathrm{mod}(x^v - 1)\,.$$

So, for all v^{th} roots of unity $\zeta \neq 1$

$$\theta^2(\zeta) + \theta(\zeta) + n = 0 \tag{4.3}$$

$$\theta(\zeta) = (-1 \pm \sqrt{-v}\,)/2 \tag{4.4}$$

since $v = 4n - 1$ in any Hadamard difference set.

Since $v \equiv 3 \pmod 4$, it has an odd prime divisor q with $q \equiv 3 \pmod 4$ for which q^s divides v but q^{s+1} does not divide v and s is an odd integer. Let $p \neq q$ be another odd prime divisor of v and let $\zeta = \zeta_p = e^{2\pi i/p}$, then equation (4.4) shows that $(-1 \pm \sqrt{-v}\,)/2$ is an algebraic integer of the field of pth roots of unity. Contradiction. So v is an odd power of the prime q, where

$q \equiv 3 \pmod 4$.

Let $\zeta_q = e^{2\pi i/q}$ and let

$$z(x) = \frac{1}{2} + x + x^4 + \cdots + x^{[(q-1)/2]^2} \qquad (4.5)$$

so that $z(\zeta_q) = \frac{1}{2}\sqrt{-q}$ (Gaussian sum). We shall use the constructive process discussed in section III.D. To do this we need the complete set of θ_d's for all divisors d of v $(=q^s)$. Equation (4.4) provides us with these, for from (4.4), (4.5) it follows that $\theta_1(x) = k = (q^s - 1)/2$ and that

$$\theta_{q^i}(x) = -\frac{1}{2} + \varepsilon_i\, q^{(s-1)/2}\, z(x^{q^{i-1}}) \quad \text{for} \quad i = 1,\ldots,s\,.$$

Here $\varepsilon_i = \pm 1$ and there is no loss of generality in assuming that $\varepsilon_1 = +1$. From (3.19), (3.20) it follows that modulo $x^v - 1$

$$\theta(x) \equiv \frac{1}{v}\sum_{w|v} \theta_w(x)\, B_{v,w}(x) \equiv \frac{v-1}{2v}\left(\frac{x^v-1}{x-1}\right)$$

$$+\ \frac{1}{v}\sum_{\substack{w|v\\ w\neq 1}} \theta_w(x) \sum_{r|w} \mu\left(\frac{w}{r}\right) r\ \frac{x^v-1}{x^r-1}\,.$$

Since $\mu(1) = 1$, $\mu(q) = -1$ and $\mu(q^i) = 0$ for $i = 2,\ldots,s$, this last summation only involves $r = w$ and $r = w/q$. So interchanging the order of summation and matching terms yields

$$\theta(x) \equiv \theta_v(x) + \frac{1}{v}\left[\frac{v-1}{2} - \theta_q(x)\right]\left(\frac{x^v-1}{x-1}\right)$$

$$+\ \frac{1}{v}\left[\sum_{i=1}^{s-1} q^i \left(\frac{x^v-1}{x^{q^i}-1}\right)\{\theta_{q^i}(x) - \theta_{q^{i+1}}(x)\}\right].$$

Now, for $i = 0,1,\ldots,s$,

$$x^{q^i}\left(\frac{x^v - 1}{x^{q^i} - 1}\right) \equiv \frac{x^v - 1}{x^{q^i} - 1} \qquad \mathrm{mod}(x^v - 1) .$$

Using this fact

$$\begin{aligned}\theta(x) \equiv \theta_v(x) + \frac{1}{2v}\left[v - q^{(s+1)/2}\right]\left(\frac{x^v - 1}{x - 1}\right) \\ + \frac{1}{v}\sum_{j=1}^{s-1}\left[q^{(s+2j-1)/2}\{\varepsilon_j z(x^{q^{j-1}}) - \varepsilon_{j+1}(q/2)\}\left(\frac{x^v - 1}{x^{q^j} - 1}\right)\right]\end{aligned} \tag{4.6}$$

so the only question is whether this has integer coefficients. If $s = 1$, i.e., if $v = q$, then (4.6) becomes ($\varepsilon_1 = 1$, as noted earlier)

$$\theta(x) \equiv -\frac{1}{2} + z(x) = -\frac{1}{2} + \frac{1}{2} + x + x^4 + \cdots + x^{[(q-1)/2]^2}$$

which does indeed have integer coefficients and in fact is the quadratic residue difference set for $q \equiv 3 \pmod 4$.

Let s be odd and $s \geq 3$ then in order for (4.6) to have integral coefficients, it is necessary that the terms

$$\frac{-q^{(s+1)/2}}{2v}\left(\frac{x^v - 1}{x - 1}\right) \quad \text{and} \quad \frac{1}{v}\, q^{(s+1)/2}\{\varepsilon_1 z(x)\}\left(\frac{x^v - 1}{x^q - 1}\right) \tag{4.7}$$

compensate for each other, since they are the only terms of (4.6) which have the coefficient

$$\frac{q^{(s+1)/2}}{v} = \frac{1}{q^{(s-1)/2}} .$$

All other terms of (4.6) contain higher powers of q. But the term on the left side of (4.7) contributes to all q^s coefficients whereas the other term of (4.7) contributes to $(q^s + q^{s-1})/2$ coefficients. So they cannot compensate for each other and therefore no such difference set exists. Thus Theorem 4.15 has been established.

Note: Any symmetric block design (i.e., not necessarily cyclic) with parameters $v,k,\lambda = 4t - 1, 2t - 1, t - 1$ is called a Hadamard design, which, of course, gives rise to the name Hadamard difference set for the associated difference set of a cyclic design. If -1 is used instead of 0 in the incidence matrix of a Hadamard design, the resultant matrix is such that the inner product of any two distinct rows is -1. Thus, by adding a constant row and column of $+1$'s to this matrix, a matrix of ± 1's is constructed whose distinct row inner products are uniformly 0. That is, a ± 1 square matrix of order $4t$ whose rows are mutually orthogonal. Such a matrix is called a Hadamard matrix because its determinant achieves the upper bound specified by Hadamard's determinantal inequality [i.e., abs. val. det. $= (4t)^{2t}$]. Thus the name Hadamard passed from the inequality to the matrices, to the associated block designs and finally to the difference sets. For example, the $v,k,\lambda = 7, 3, 1$ Hadamard difference set $\{1,2,4\}$ has associated modified incidence matrix and Hadamard matrix:

$$\begin{matrix}
-1 & 1 & 1 & -1 & 1 & -1 & -1 \\
-1 & -1 & 1 & 1 & -1 & 1 & -1 \\
-1 & -1 & -1 & 1 & 1 & -1 & 1 \\
1 & -1 & -1 & -1 & 1 & 1 & -1 \\
-1 & 1 & -1 & -1 & -1 & 1 & 1 \\
1 & -1 & 1 & -1 & -1 & -1 & 1 \\
1 & 1 & -1 & 1 & -1 & -1 & -1
\end{matrix}
\qquad
\begin{matrix}
1 & 1 & 1 & 1 & 1 & 1 & 1 & 1 \\
1 & -1 & 1 & 1 & -1 & 1 & -1 & -1 \\
1 & -1 & -1 & 1 & 1 & -1 & 1 & -1 \\
1 & -1 & -1 & -1 & 1 & 1 & -1 & 1 \\
1 & 1 & -1 & -1 & -1 & 1 & 1 & -1 \\
1 & -1 & 1 & -1 & -1 & -1 & 1 & 1 \\
1 & 1 & -1 & 1 & -1 & -1 & -1 & 1 \\
1 & 1 & 1 & -1 & 1 & -1 & -1 & -1
\end{matrix}$$

Chapter 14 of Hall (1967) surveys (with proofs) most of the known results on the existence of Hadamard matrices; for results subsequent to Hall's book see Spence (1967), Goethals and Seidel (1967), Wallis (1969 AB, 1970), Turyn (1970) and Whiteman (1970). It is known that the order of such a matrix is 1, 2 or $4t$ and the first few undecided cases are $n = 188, 236, 260, 268, 292$.

C. Barker Sequences, Circulant Hadamard Matrices

In 1953 R. H. Barker (1953), in connection with a problem in digital communications, considered the question of the existence of finite sequences of ones and minus ones $\{b_i\}_1^v$, with the property that their aperiodic auto-correlation coefficients should be as small as possible. That is, he asked that

$$c_j = \sum_{i=1}^{v-j} b_i \, b_{i+j} = 0 \quad \text{or} \quad -1$$

for all j, $1 \leq j \leq v - 1$. He found such sequences for $v = 3, 7, 11$. It has become customary to relax Barker's condition slightly and call all finite 1, -1 sequences, whose aperiodic autocorrelations c_j are restricted to -1, 0, 1, Barker sequences. Only the following Barker sequences are known: (+ denotes +1 and - denotes -1)

$v = 2$ + +

$v = 3$ + + -

$v = 4$ + + + - ; + + - +

$v = 5$ + + + - +

$v = 7$ + + + - - + -

$v = 11$ + + + - - - + - - + -

$v = 13$ + + + + + - - + + - + - +

together with the sequences which may be derived from them by the following transformations:

$$b_i' = (-1)^i \, b_i$$

$$b_i' = (-1)^{i+1} \, b_i$$

$$b_i' = -b_i \, .$$

In fact, Storer and Turyn (1961) have shown that any further Barker sequences which may exist must be of even length, indeed they show that $v \equiv 0 \pmod 4$ is necessary.

Note that in terms of the 1, -1 representation of the characteristic function of a difference set (see section I.D. for this) all these sequences correspond to difference sets. For $v = 2,3,4,5$ the difference sets are trivial; for $v = 7,11,13$ the sets have parameters $v,k,\lambda = 7, 4, 2;\ 11, 5, 2$ and $13, 9, 6$ respectively.

It can be shown (Storer and Turyn, 1961) that $c_j + c_{v-j} \equiv v \pmod 4$ in any Barker sequence. Further, if $v \equiv 0 \pmod 4$ then $c_j + c_{v-j} = 0$. Thus, as $c_j + c_{v-j} = R_b(j)$ [the correlation coefficient defined in section I.D], any further Barker sequence $\{b_i\}$ has autocorrelation function

$$R_b(j) = \begin{cases} v & \text{if } j \equiv 0 \text{ modulo } v \\ 0 & \text{otherwise} \end{cases}$$

i.e., a two-level autocorrelation function. Thus such a sequence corresponds to a difference set. Since v is even, $n = k - \lambda$ is a square (say N^2) by Theorem 2.1 and since $R_b(j) = v - 4(k - \lambda) = 0$ for $j \not\equiv 0$ modulo v it follows that $v = 4N^2$. Since $k(k - 1) = \lambda(v - 1)$ for any difference set and $n = k - \lambda$ these values of n, v show that $k = 2N^2 - N$ or $2N^2 + N$. Since these k's correspond to complementary difference sets, there is no loss of generality in assuming that $v,k,\lambda = 4N^2,\ 2N^2 - N,\ N^2 - N$.

Thus, further Barker sequences exist if and only if there exist difference sets with parameters $v,k,\lambda = 4N^2,\ 2N^2 - N,\ N^2 - N$ for $N > 1$. Since $(v,n) = (4N^2,N^2) = N^2 > 1$ here, this is a subcase of the unsolved problem concerning the existence of cyclic difference sets with $(v,n) > 1$. Theorems 2.13 and 2.17 above rule out many of these cases; in particular, all the cases $1 < N < 55$ with the single exception $N = 39$. Turyn (1968) excludes this case by a constructive method; essentially that of section III.D. Thus, if any further Barker sequences exist they must have $N \geq 55$, i.e., $v \geq 12{,}100$.

A matrix is called <u>circulatory</u> or said to be a <u>circulant</u> if each successive row is derived from the previous row by shifting it cyclically one position <u>to the</u>

right. For example, the matrix (+ for +1, - for -1)

+	+	+	-
-	+	+	+
+	-	+	+
+	+	-	+

is a circulant. This particular circulant has every entry ± 1 and its rows are orthogonal to each other. Matrices with these properties are called Hadamard matrices and they have been extensively studied [see the note at the end of section IV.B]. The example, thus, is a circulant Hadamard matrix. In fact, up to rearrangement and scalar multiplication by -1 of its rows and columns, it is the only known circulant Hadamard matrix. It follows immediately from the autocorrelation function of a Barker sequence of even length $v \geq 4$, that there is a one-to-one correspondence between such sequences and circulant Hadamard matrices. Thus (from the Barker sequence results above), if there exists any further circulant Hadamard matrices they have orders $v \geq 12{,}100$. It should come as no surprise then, that the absence of any further Barker sequences/circulant Hadamard matrices has been conjectured. [Using the fact that -1 is never a multiplier of a non-trivial cyclic difference set (Theorem 3.3 above), Brualdi (1965) has shown that there does not exist an Hadamard matrix of order $v > 4$ which is a symmetric circulant.]

The related problem of finding $1, -1$ sequences of length v for which the maximum aperiodic correlation coefficient is of least magnitude (i.e., for which

$$\max_j |c_j|$$

is minimized) and indeed the problem of determining this minimum, at least asymptotically as a function of v, is unsolved. Turyn (1968) provides a survey of the known results on this subject.

V. FAMILIES OF DIFFERENCE SETS

The known difference sets (with a few exceptions) can be divided into families. This chapter deals with these families of difference sets, construction methods specific to them, their multiplier groups and the status of some open questions related to them.

A. Singer Sets and Their Generalizations. The Results of Gordon, Mills and Welch

Singer (1938) discovered a large class of difference sets related to finite projective geometries. These have parameters:

$$v = \frac{q^{N+1} - 1}{q - 1}, \quad k = \frac{q^N - 1}{q - 1}, \quad \lambda = \frac{q^{N-1} - 1}{q - 1} \tag{5.1}$$

for $N \geq 1$ and they exist whenever q is a prime power.

In order to discuss Singer's result properly it is necessary to know some of the theory of finite fields [see, for example, van der Waerden (1949) for proofs]. For any prime power q there exists a finite field with exactly q elements. This field is unique up to isomorphism and is called <u>the Galois field of q elements</u> [written $GF(q)$]. The multiplicative group of $GF(q)$ is cyclic; thus it is generated by any of its $\varphi(q-1)$ elements of order $q-1$. These generating elements are called <u>primitive roots</u> and if α is a primitive root so is α^u whenever u is prime to $q-1$. For prime p, the residues $0,1,\ldots,p-1$ form a field with respect to addition and multiplication modulo p; this field is often taken to be the generic representation of $GF(p)$. $GF(r)$, $r = q^m$, can be constructed from $GF(q)$ by adjoining any root β of any m^{th} degree polynomial $f(x)$ irreducible over $GF(q)$. The subfields of $GF(p^m)$, p prime, are $GF(p^j)$ for all divisors j of m.

$GF(q^m)$ is often represented by the set of all m-tuples with entries from $GF(q)$. In this representation addition is performed componentwise but multiplication is more complicated. Associate with the m-tuple $a_{m-1}, a_{m-2}, \ldots, a_1, a_0$ the polynomial $a_{m-1} x^{m-1} + \cdots + a_1 x + a_0$. Then, in order to multiply two m-tuples, multiply instead their associated polynomials and reduce the result modulo any <u>fixed</u> m^{th} degree polynomial $f(x)$ irreducible over $GF(q)$. The coefficients of the resulting polynomial constitute the m-tuple which is the product of the original two. For multiplicative purposes it is more convenient to represent $GF(q^m)$ in terms of a primitive root α; in which case, $GF(q^m)$ consists of

$$0, 1, \alpha, \alpha^2, \ldots, \alpha^{q^m - 2} .$$

Multiplication then becomes a simple matter of reducing exponents modulo $q^m - 1$ but addition is more complicated. Both these representations of $GF(q^m)$ are used in the proof of Singer's theorem. Table 5.1 shows both kinds of representation for

TABLE 5.1. $GF(2^6)$ with $f(x) = x^6 + x + 1$.

0	0 0 0 0 0 0	α^{15}	1 0 1 0 0 0	α^{31}	1 0 0 1 0 1	α^{47}	1 0 0 1 1 1
1	0 0 0 0 0 1	α^{16}	0 1 0 0 1 1	α^{32}	0 0 1 0 0 1	α^{48}	0 0 1 1 0 1
α	0 0 0 0 1 0	α^{17}	1 0 0 1 1 0	α^{33}	0 1 0 0 1 0	α^{49}	0 1 1 0 1 0
α^2	0 0 0 1 0 0	α^{18}	0 0 1 1 1 1	α^{34}	1 0 0 1 0 0	α^{50}	1 1 0 1 0 0
α^3	0 0 1 0 0 0	α^{19}	0 1 1 1 1 0	α^{35}	0 0 1 0 1 1	α^{51}	1 0 1 0 1 1
α^4	0 1 0 0 0 0	α^{20}	1 1 1 1 0 0	α^{36}	0 1 0 1 1 0	α^{52}	0 1 0 1 0 1
α^5	1 0 0 0 0 0	α^{21}	1 1 1 0 1 1	α^{37}	1 0 1 1 0 0	α^{53}	1 0 1 0 1 0
α^6	0 0 0 0 1 1	α^{22}	1 1 0 1 0 1	α^{38}	0 1 1 0 1 1	α^{54}	0 1 0 1 1 1
α^7	0 0 0 1 1 0	α^{23}	1 0 1 0 0 1	α^{39}	1 1 0 1 1 0	α^{55}	1 0 1 1 1 0
α^8	0 0 1 1 0 0	α^{24}	0 1 0 0 0 1	α^{40}	1 0 1 1 1 1	α^{56}	0 1 1 1 1 1
α^9	0 1 1 0 0 0	α^{25}	1 0 0 0 1 0	α^{41}	0 1 1 1 0 1	α^{57}	1 1 1 1 1 0
α^{10}	1 1 0 0 0 0	α^{26}	0 0 0 1 1 1	α^{42}	1 1 1 0 1 0	α^{58}	1 1 1 1 1 1
α^{11}	1 0 0 0 1 1	α^{27}	0 0 1 1 1 0	α^{43}	1 1 0 1 1 1	α^{59}	1 1 1 1 0 1
α^{12}	0 0 0 1 0 1	α^{28}	0 1 1 1 0 0	α^{44}	1 0 1 1 0 1	α^{60}	1 1 1 0 0 1
α^{13}	0 0 1 0 1 0	α^{29}	1 1 1 0 0 0	α^{45}	0 1 1 0 0 1	α^{61}	1 1 0 0 0 1
α^{14}	0 1 0 1 0 0	α^{30}	1 1 0 0 1 1	α^{46}	1 1 0 0 1 0	α^{62}	1 0 0 0 0 1

$GF(2^6)$. [In this example the primitive root α satisfies $f(x) = 0$. This is by no means necessary; however it is often quite convenient and it is theoretically always possible to arrange things this way. That is, a primitive root of $GF(q^m)$ will always satisfy an irreducible polynomial of degree m over $GF(q)$; such a polynomial is said to be primitive of degree m over $GF(q)$. The trick is to find such polynomials. Alanen and Knuth (1964, Table 7) list, for each prime $p < 50$ and all exponents m such that $p^m < 10^9$, a primitive irreducible polynomial of degree m over $GF(p)$.]

The finite projective geometry $PG(N,q)$, of dimension N over $GF(q)$, consists of all $(N + 1)$-tuples of elements of $GF(q)$ subject to the restriction that $(a_N, a_{N-1}, \ldots, a_0)$ and $(ba_N, ba_{N-1}, \ldots, ba_0)$ are identified for all $b \neq 0$, b in $GF(q)$. Thus any set of $j + 1$ linearly independent $(N + 1)$-tuples determines a subspace of $PG(N,q)$ of dimension j. The $(q^{N+1} - 1)/(q - 1)$ subspaces of dimension 0 are the points of this geometry and the $(q^{N+1} - 1)/(q - 1)$ subspaces of dimension $N - 1$ are called hyperplanes. Any two distinct hyperplanes intersect in a subspace of dimension $N - 2$; so they have $(q^{N-1} - 1)/(q - 1)$ points in common. Thus there are $v = (q^{N+1} - 1)/(q - 1)$ points and v hyperplanes in $PG(N,q)$ and $k = (q^N - 1)/(q - 1)$ points in any hyperplane. Singer (1938) has shown:

Theorem 5.1. Considering the points as objects and the hyperplanes as blocks, $PG(N,q)$ forms a symmetric block design with parameters given by (5.1) above. This block design is cyclic; thus the points of any hyperplane determine a v, k, λ difference set.

Proof. The discussion above shows that A, the incidence matrix of this configuration, satisfies

$$AA^T = (k - \lambda)I + \lambda J, \qquad AJ = kJ .$$

Thus conditions (ii), (iii) of the block design definition follow from the result of Ryser (1950) discussed in section II.C. just below Theorem 2.3. So $PG(N,q)$

forms a symmetric block design as indicated.

It remains to be shown that there exists a numbering of the points and hyperplanes of $PG(N,q)$ which demonstrates the cyclic nature of the design. Let α be a primitive root of $GF(q^{N+1})$ and let α satisfy the irreducible polynomial

$$f(x) = x^{N+1} + c_N x^N + \cdots + c_1 x + c_0 \tag{5.2}$$

over $GF(q)$. Then each power of α corresponds to a unique $(N + 1)$-tuple over $GF(q)$. Since α^{vi} belongs to $GF(q)$ for all i, the elements of $GF(q)$ are $0, 1, \alpha^v, \ldots, \alpha^{(q-2)v}$ and it follows that α^j and α^{j+vi} correspond to the same point of $PG(N,q)$. Thus there is a one-to-one correspondence between the elements $\alpha^j (j = 0,1,\ldots,v - 1)$ and the points of $PG(N,q)$ which assigns to every point of $PG(N,q)$ an exponent j, $0 \leq j \leq v - 1$.

Consider the mapping σ

$$\sigma : \alpha^i \to \alpha^{i+1} \qquad \sigma : 0 \to 0 \tag{5.3}$$

or in additive notation, using equation (5.2),

$$\sigma : (a_N, \ldots, a_1, a_0) \to (a_{N-1} - a_N c_N, \ldots, a_0 - a_N c_1, - a_N c_0) \,. \tag{5.4}$$

This mapping obviously maps points onto points and [as is clear from (5.4)] maps subspaces onto subspaces without any loss of dimension. Thus, it maps hyperplanes into hyperplanes. Since the point corresponding to α^{j+iv} is the same as that corresponding to α^j, the mapping σ is cyclic of order v on the points. If σ is not cyclic of order v on the hyperplanes then there exists a hyperplane H and an integer s, $1 \leq s \leq v - 1$, such that σ^s fixes H. Let α^i be a point of H, then $\alpha^i, \alpha^{i+s}, \ldots, \alpha^{i+ts} = \alpha^i$ form an orbit in H under σ^s with $\alpha^i(\alpha^{ts} - 1) = 0$. So t is necessarily the least positive integer such that v divides ts. This is, of course, independent of which element α^i of H was chosen; so t divides k. Since v does not divide s (by assumption) the fact that v divides ts implies that $(v,t) > 1$ and as t divides k it follows that $(v,k) > 1$; this contradicts the fact that $v - qk = 1$. Thus v divides s necessarily and σ is cyclic of order v on the hyperplanes. So the theorem of Singer has been established. Thus, if one lists the elements of any hyperplane (say $\alpha^i, \alpha^j, \ldots, \alpha^m$), their exponents form a difference set with parameters given by (5.1) above.

Any mapping L from $GF(q^{N+1})$ onto $GF(q)$ which satisfies

$$L(b\beta + c\gamma) = bL(\beta) + cL(\gamma)$$

for all β, γ in $GF(q^{N+1})$ and all b, c in $GF(q)$ is called a <u>linear functional</u> from $GF(q^{N+1})$ to $GF(q)$. The set of elements β of $GF(q^{N+1})$ annihilated by such a linear functional L [i.e., such that $L(\beta) = 0$] constitutes a hyperplane in $PG(N,q)$. Further, every hyperplane of $PG(N,q)$ is annihilated by some linear functional from $GF(q^{N+1})$ to $GF(q)$. Thus to apply Singer's Theorem one merely computes the null space of a single linear functional. For example, when $q = 2$ and $N = 5$ then v, k, λ are 63, 31, 15 and α is to be a root of a primitive 6th degree polynomial (say $x^6 + x + 1 = 0$) over $GF(2)$. Consider the linear functional L which maps each element into its right-most component. Table 5.1 shows that L annihilates α^i when i is one of 1, 2, 3, 4, 5, 7, 8, 9, 10, 13, 14, 15, 17, 19, 20, 25, 27, 28, 29, 33, 34, 36, 37, 39, 42, 46, 49, 50, 53, 55, 57. Thus these numbers constitute a difference set with parameters $v,k,\lambda = 63,31,15$.

Note: Singer's construction can be varied in several inessential ways, the only effect of this is to generate a difference set equivalent to the original one. For example, if a different hyperplane (or linear functional) is used this merely shifts the set. If the primitive root α^t, $tr \equiv 1 \bmod (q^{N+1} - 1)$, is used instead of α, the equivalent set $rD = \{rd_1, \ldots, rd_k\}$ results. In fact not even a primitive root is needed. It is only necessary to have an element $\beta (= \alpha^u)$ for which β^i is <u>not</u> in $GF(q)$ for $i = 1, \ldots, v - 1$. [The use of β results in the same difference set as that generated by the primitive root α^{u+jv}, the existence of which (for some value of j) is guaranteed by the condition on β.]

Extensive use will be made of the following well-known fact:

Lemma 5.2. Let L be any linear functional, not identically zero, from $GF(q^m)$ to its subfield $GF(q^j)$ and let μ be any element of $GF(q^m)$. Then every linear functional from $GF(q^m)$ to $GF(q^j)$ is of the form L_μ, where L_μ is defined by $L_\mu(\beta) = L(\mu\beta)$ for all β in $GF(q^m)$. Moreover if $\mu \neq \nu$ then $L_\mu \neq L_\nu$.

[Linear algebra provides the fact that there are precisely q^m linear functionals from $GF(q^m)$ to $GF(q^j)$ and the above process constructs q^m distinct ones.]

Complementary to any Singer set D is a difference set D^* with parameters

$$v = \frac{q^{N+1} - 1}{q - 1}, \qquad k = q^N, \qquad \lambda = q^{N-1}(q-1) \tag{5.5}$$

for $N \geq 1$ and q a prime power. Here $j = 0,1,\ldots,v-1$ belongs to D^* if and only if $L(\alpha^j) \neq 0$ where L is a fixed linear functional and α is a primitive root of $GF(q^{N+1})$. Call this difference set $D(L,\alpha)$. Note that there is no loss of generality in assuming that $L(1) = 1$, for this may be arranged by taking a different linear functional.

The difference set $D(L,\alpha)$ corresponding to the example given above consists of 0, 6, 11, 12, 16, 18, 21, 22, 23, 24, 26, 30, 31, 32, 35, 38, 40, 41, 43, 44, 45, 47, 48, 51, 52, 54, 56, 58, 59, 60, 61, 62. Examination of these residues shows that modulo 9 they constitute 4 copies of the trivial difference set $E = \{0,2,3,4,5,6,7,8\}$. In fact $D(L,\alpha)$ gives rise to the array of Table 5.2

TABLE 5.2. A Representation of $D(L,\alpha)$.

j	0	1	2	3	4	5	6
i = 0	1	0	1	0	0	1	1
1	0	0	0	0	0	0	0
2	0	1	0	0	1	1	1
3	0	1	1	1	0	1	0
4	0	0	1	1	1	0	1
5	0	0	1	1	1	0	1
6	1	0	1	0	0	1	1
7	0	1	0	0	1	1	1
8	0	0	1	1	1	0	1

where the (i,j) entry is 1 if and only if $i + 9j$ belongs to $D(L,\alpha)$. Note that, for each i in the difference set E, the rows of this array are the characteristic function of the $w,\ell,\mu = 7,4,2$ difference set $F = \{0,2,5,6\}$ or one of its shifts. Gordon, Mills and Welch (1962) have shown that the structure of $D(L,\alpha)$ always depends upon difference sets E and F in this manner. Specifically, they prove:

Theorem 5.3. Let q be a power of the prime p and let $N + 1$ be an integer, $N \geq 1$. Let L be a linear functional from the finite field $GF(q^{N+1})$ to the subfield $GF(q)$, such that $L(1) = 1$. Let L_0 be the restriction of L to an intermediate field $GF(q^m)$, where m divides $N + 1$. Let $\tilde{L}$ be the linear functional which assigns to each ζ in $GF(q^{N+1})$ the unique element $\tilde{L}(\zeta)$ in $GF(q^m)$, which satisfies the relation $L_0(\tilde{L}(\zeta)\delta) = L(\zeta\delta)$ for all δ in $GF(q^m)$. Set $v = (q^{N+1} - 1)/(q-1)$, $w = (q^m - 1)/(q-1)$ and $\xi = v/w$. Let α be a primitive root of $GF(q^{N+1})$ and let $\beta = \alpha^{\xi}$. Let $\theta(x)$ and $\psi(y)$ be the Hall polynomials of $D(L,\alpha)$ and $D(L_0,\beta)$ respectively. [For $m = 1$, take $\psi(y) = 1$.] Let $y = x^{\xi}$. Then

$$\theta(x) \equiv \Omega(x)\,\psi(y) \qquad (\text{mod } x^v - 1) \tag{5.6}$$

where

$$\Omega(x) = \Sigma\, x^i y^{r_i} \tag{5.7}$$

and this summation is taken over those values of i for which

$$\tilde{L}(\alpha^i) \neq 0\,, \quad 0 \leq i < \xi\,, \quad \text{and} \quad \tilde{L}(\alpha^i) = \beta^{-r_i}\,. \tag{5.8}$$

In the example above $N + 1, m, q, p = 6, 3, 2, 2$, $F = D(L_0, \beta) = \{0,2,5,6\}$ while E is the difference set determined by $\tilde{L}$ and the extension $GF(q^{N+1})$ of $GF(q^m)$. The r_i determine the shifts associated with the various copies of F in the array.

Proof. Consider $\tilde{L}$. Let ζ be an element of $GF(q^{N+1})$. Then the mapping $\delta \to L(\zeta\delta)$, δ in $GF(q^m)$, is a linear functional from $GF(q^m)$ to $GF(q)$. So, by Lemma 5.2, there is a unique element, call it $\tilde{L}(\zeta)$, of $GF(q^m)$, such that

$$L_0(\tilde{L}(\zeta)\delta) = L(\zeta\delta) \tag{5.9}$$

for all δ in $GF(q^m)$. [L_0 is not identically zero since $L(1) = 1$.] Thus $\tilde{L}$ is a properly defined mapping from $GF(q^{N+1})$ to $GF(q^m)$. The only question which remains is whether or not $\tilde{L}$ is a linear functional. Let a, b, δ be elements of $GF(q^m)$, a and b fixed, and let ζ, η be fixed elements of $GF(q^{N+1})$ then

$$\begin{aligned} L_0(\tilde{L}(a\eta + b\zeta)\delta) &= L((a\eta + b\zeta)\delta) = L(a\eta\delta) + L(b\zeta\delta) \\ &= L_0(\tilde{L}(\eta)a\delta) + L_0(\tilde{L}(\zeta)b\delta) \\ &= L_0((a\tilde{L}(\eta) + b\tilde{L}(\zeta))\delta)\,. \end{aligned}$$

So $\tilde{L}$ is indeed a linear functional and in fact $\tilde{L}(1) = 1$. When $m = 1$, it follows from $L_0(1) = 1$ that L_0 is the identity mapping on $GF(q)$; since δ belongs to $GF(q)$ when $m = 1$, it follows from (5.9) that $\tilde{L} = L$ in this case. Let

$$\psi(y) = \sum_{j=0}^{w-1} \delta_j y^j \quad \text{with} \quad \delta_j = \begin{cases} 0 & \text{if } L_0(\beta^j) = 0 \\ 1 & \text{if } L_0(\beta^j) \neq 0 \end{cases}$$

and let

$$\theta(x) = \sum_{i=0}^{v-1} \varepsilon_i x^i \quad \text{with} \quad \varepsilon_i = \begin{cases} 0 & \text{if } L(\alpha^i) = 0 \\ 1 & \text{if } L(\alpha^i) \neq 0 \end{cases}$$

$$= \sum_{i=0}^{\xi-1} x^i \omega_i(y) \quad \text{when} \quad \omega_i(y) = \sum_{j=0}^{w-1} \varepsilon_{i+\xi j} y^j .$$

Since β is a primitive root of $GF(q^m)$ every value of $\tilde{L}(\alpha^i)$ is either 0 or a power of β, say β^{-r_i}. Now $\varepsilon_{i+\xi j} = 0$ if and only if $L(\alpha^{i+\xi j}) = 0$ and

$$L(\alpha^{i+\xi j}) = L(\alpha^i \beta^j) = L_0(\tilde{L}(\alpha^i)\beta^j) = \begin{cases} 0 & \text{if } \tilde{L}(\alpha^i) = 0 \\ L_0(\beta^{j-r_i}) & \text{if } \tilde{L}(\alpha^i) \neq 0 . \end{cases}$$

So

$$\varepsilon_{i+\xi j} = \begin{cases} 0 & \text{if } \tilde{L}(\alpha^i) = 0 \\ \delta_{j-r_i} & \text{if } \tilde{L}(\alpha^i) \neq 0 . \end{cases}$$

Thus $\omega_i(y) = 0$ if $\tilde{L}(\alpha^i) = 0$, while if $\tilde{L}(\alpha^i) \neq 0$

$$\omega_i(y) = \sum_{j=0}^{w-1} \varepsilon_{i+\xi j}\, y^j = \sum_{j=0}^{w-1} \delta_{j-r_i}\, y^j \equiv \sum_{j=0}^{w-1} \delta_j\, y^{j+r_i} = y^{r_i}\, \psi(y) \quad (\text{mod } x^v - 1) .$$

Hence $\theta(x) \equiv \psi(y)\, \Sigma\, x^i y^{r_i}$, where the sum is taken as prescribed by (5.8), and the theorem is proved.

Now $\psi(y)$ is the Hall polynomial of the w,ℓ,μ-difference set $D(L_0,\beta)$ which has parameter values (for $m > 1$)

$$w = \frac{q^m - 1}{q - 1}, \qquad \ell = q^{m-1}, \qquad \mu = q^{m-2}(q - 1). \tag{5.10}$$

Gordon, Mills and Welch show that, if $\psi(y)$ is replaced in (5.6) by the Hall polynomial $\psi_0(y)$ of an arbitrary difference set having parameter values (5.10), i.e., if

$$\theta_0(x) \equiv \Omega(x)\ \psi_0(y) \qquad (\mathrm{mod}\ \ x^v - 1) \tag{5.11}$$

then $\theta_0(x)$ is again the Hall polynomial of a v,k,λ-difference set with parameters (5.5). For let

$$\psi_0(y) = \sum_{j=0}^{w-1} \eta_j y^j$$

be a Hall polynomial and let $y = x^\xi$. Let $\Omega(x)$ be given by (5.7). Then

$$\theta_0(x) = \Omega(x)\ \psi_0(y) = \sum_{i=1}^{k} x^{e_i}$$

where the e_i are distinct modulo v by construction. Further, by definition, modulo $x^v - 1$,

$$\psi_0(y)\ \psi_0(y^{-1}) \equiv (\ell - \mu) + \mu(1 + y + \dots + y^{w-1}) \equiv \psi(y)\ \psi(y^{-1}) .$$

Hence, modulo $x^v - 1$,

$$\theta_0(x)\ \theta_0(x^{-1}) = \Omega(x)\ \Omega(x^{-1})\ \psi_0(y)\ \psi_0(y^{-1}) \equiv \Omega(x)\ \Omega(x^{-1})\ \psi(y)\ \psi(y^{-1})$$

$$= \theta(x)\ \theta(x^{-1}) \ .$$

Thus $e_1,\ldots,e_k$ form a difference set with the parameter values of (5.5). [For example, let $\psi_0(y) = 1 + y + y^2 + y^5$ then using

$$\Omega(x) = 1 + x^2y^6 + x^3y^3 + x^4y^4 + x^5y^4 + x^6 + x^7y^6 + x^8y^4$$

from $D(L,\alpha)$ above, in congruence 5.11 yields the $v,k,\lambda = 63,32,16$ difference set 0, 2, 6, 7, 9, 11, 12, 15, 16, 18, 22, 23, 24, 26, 30, 38, 39, 40, 41, 43, 44, 45, 48, 49, 50, 51, 53, 56, 58, 59, 61, 62. This difference set is not equivalent to $D(L,\alpha)$ as is explained below.]

Gordon, Mills and Welch show that two v,k,λ-difference sets derived from (5.11), are equivalent if and only if $\psi_0(y) = y^s\ \psi(y)$. That is, if and only if the w,ℓ,μ-difference sets are shifts of each other. [Since -1 is never a multiplier of a cyclic difference set (section 3.1) the difference sets D and $-D$ are never shifts of each other. Thus the polynomials $\psi(y)$ and $\psi(y^{-1})$ always generate inequivalent v,k,λ-difference sets by these means. This accounts for the inequivalence of the two 63, 32, 16 difference sets mentioned above.] Several lemmas are required for the proof of this result (which is Theorem 5.12 below).

Following Gordon, Mills and Welch, let $B = \{b_1,b_2,\ldots,b_\ell\}$ and $C = \{c_1,c_2,\ldots,c_\ell\}$ be two w,ℓ,μ-difference sets, and let

$$\psi_b(y) = \sum_i y^{b_i}, \quad \psi_c(y) = \sum_i y^{c_i}$$

be their Hall polynomials. [If $m = 1$ let $\psi_b(y) = \psi_c(y) = 1$.] Put

$$\theta_b(x) = \Omega(x)\,\psi_b(y), \qquad \theta_c(x) = \Omega(x)\,\psi_c(y).$$

Then $\theta_b(x)$ and $\theta_c(x)$ are the Hall polynomials of two v,k,λ-difference sets, say $\overline{B}$ and $\overline{C}$. If $\overline{B}$ and $\overline{C}$ are equivalent then there exist integers a and t such that $(t,v) = 1$ and

$$\theta_b(x) \equiv x^a \theta_c(x^t) \qquad (\mathrm{mod}\ x^v - 1). \tag{5.12}$$

Lemma 5.4. If (5.12) holds, then there exists integers r and s such that

$$\Omega(x) \equiv x^r\,\Omega(x^t) \qquad (\mathrm{mod}\ x^v - 1) \tag{5.13}$$

$$\psi_b(y) \equiv y^s\,\psi_c(y^t) \qquad (\mathrm{mod}\ y^w - 1).$$

In particular if B and C are inequivalent then so are $\overline{B}$ and $\overline{C}$.

Proof. By construction

$$\theta_b(x) = \Sigma\, x^i y^{r_i}\,\psi_b(y) \quad \text{and} \quad x^a\,\theta_c(x) = \Sigma\, x^{a+ti} y^{tr_i}\,\psi_c(y^t)$$

where these summations are taken over those values of i for which $\tilde{L}(\alpha^i) \neq 0$, $0 \leq i < \xi$. Let j be such an i; then, comparing terms in (5.12)

$$x^i y^{r_i}\,\psi_b(y) \equiv x^{a+th} y^{tr_h}\,\psi_c(y^t) \qquad (\mathrm{mod}\ x^v - 1)$$

where $a + th \equiv j$ modulo ξ. Since $x^v - 1 = y^w - 1$ this yields

$$\psi_b(y) \equiv y^s\,\psi_c(y^t) \qquad (\mathrm{mod}\ y^w - 1) \tag{5.14}$$

where $s = tr_h - r_j + \xi^{-1}(a + th - j)$.

Now $\psi_b(y)\,\psi_b(y^{-1}) \equiv \ell - \mu + \mu(1 + y + \cdots + y^{w-1})$ modulo $y^w - 1$ and this implies that $\psi_b(y)$ is relatively prime to $y^w - 1$. So from (5.12), i.e., from

$$\Omega(x)\,\psi_b(y) \equiv x^a \Omega(x^t)\,\psi_c(y^t) \qquad (\text{mod } x^v - 1)$$

and (5.14)

$$\Omega(x) \equiv x^a y^{-s}\,\Omega(x^t) \qquad (\text{mod } x^v - 1)$$

and the lemma is proved.

Let $Q = GF(q)$ and let Q^* be the set of all non-zero elements of Q.

Lemma 5.5. Suppose (5.13) holds and $(t,v) = 1$. Let $\eta = \alpha^r$, and let ω be an element of $GF(q^{N+1})$. Then $\tilde{L}(\omega)$ belongs to Q^* if and only if $\tilde{L}(\eta\omega^t)$ does also.

Proof. Since $\tilde{L}$ is a linear functional from $GF(q^{N+1})$ to $GF(q^m)$ it follows that

$$\Omega(x) = \sum x^i y^{r_i} = \sum x^{i+\xi r_i} = \sum_{j \text{ in } S} x^j \tag{5.15}$$

where S is the set of all j such that $\tilde{L}(\alpha^j) = 1$, $0 \leq j < q^N - 1$. Since α^v is a primitive root of Q^*, the effect of adding v to j is to multiply $\tilde{L}(\alpha^j)$ by a primitive root of Q^*. Thus (5.15) can be written as

$$\Omega(x) \equiv \sum_{j \text{ in } S'} x^j \qquad (\text{mod } x^v - 1)$$

where S' is the set of all j such that $\tilde{L}(\alpha^j)$ belongs to Q^*, $0 < j < v$. By (5.13) $\tilde{L}(\alpha^j)$ belongs to Q^* if and only if $\tilde{L}(\alpha^{r+jt})$ does also. If $\omega = 0$ the lemma is trivial. If $\omega \neq 0$ the result follows by putting $\omega = \alpha^j$.

Lemma 5.6. Suppose (5.13) holds and $(t,r) = 1$. Let $\eta = \alpha^r$, let ζ be an element of $GF(q^m)$ and let ω belong to $GF(q^{N+1})$. Then $\tilde{L}(\omega)$ belongs to ζQ^* if and only if $\tilde{L}(\eta\omega^t)$ belongs to $\zeta^t Q^*$.

Proof. Let $\zeta \neq 0$. Then $\tilde{L}(\omega)$ belongs to ζQ^* if and only if $\tilde{L}(\omega\zeta^{-1})$ belongs to Q^*. By Lemma 5.5 this is true if and only if $\tilde{L}(\eta\omega^t\zeta^{-t})$ belongs to Q^*, which is equivalent to $\tilde{L}(\eta\omega^t)$ being a member of $\zeta^t Q^*$. Next suppose $\tilde{L}(\omega) = 0$. Here $\tilde{L}(\eta\omega^t)$ belongs to $\nu^t Q^*$ for some ν in $GF(q^m)$. If $\nu \neq 0$, then by the first part of this proof $\tilde{L}(\omega)$ is an element of νQ^*, contradiction. So $\nu = 0$ and the proof is complete.

Lemma 5.7. Suppose that (5.13) holds and $(t,v) = 1$. Let $\zeta_1, \zeta_2, \ldots, \zeta_s$ be elements of $GF(q^{N+1})$ which are linearly independent over $GF(q^m)$. Let c_i, a_i ($1 \leq i \leq s$) be elements of $GF(q^m)$ such that

$$\left(\sum_{i=1}^{s} c_i \zeta_i \right)^t = \sum_{i=1}^{s} a_i \zeta_i^t . \tag{5.16}$$

Then a_i belongs to $c_i^t Q^*$, $1 \leq i \leq s$.

Proof. Since the ζ_i are linearly independent over $GF(q^m)$ there exist linear functionals K_j over $GF(q^m)$ such that ($1 \leq i, j \leq s$)

$$K_j(\zeta_i) = \delta_{ij} = \begin{cases} 0 & \text{if } i \neq j \\ 1 & \text{if } i = j . \end{cases}$$

By Lemma 5.2 these K_j can all be expressed in terms of $\tilde{L}$, that is there exists elements u_j of $GF(q^{N+1})$ such that

$$\tilde{L}(u_j \zeta_i) = \delta_{ij} .$$

Then, by Lemma 5.6, $\tilde{L}(\eta u_j^t \zeta_i^t) = 0$ if $i \neq j$, and $\tilde{L}(\eta u_j^t \zeta_j^t)$ belongs to Q^*. Now $\tilde{L}(u_j \Sigma c_i \zeta_i) = c_j$, so that

$$\tilde{L}(\eta u_j^t (\Sigma c_i \zeta_i)^t) \quad \text{belongs to} \quad c_j^t Q^* .$$

On the other hand, using 5.16,

$$\tilde{L}(\eta u_j^t (\Sigma c_i \zeta_i)^t) = \sum_i a_i \tilde{L}(\eta u_j^t \zeta_i^t)$$

and thus belongs to $a_j Q^*$. So a_j is an element of $c_j^t Q^*$ as was to be proved.

__Lemma 5.8.__ Suppose (5.13) holds with $(t,v) = 1$. Let $\zeta_1, \zeta_2, \ldots, \zeta_s$ be a basis for $GF(q^{N+1})$ over $GF(q^m)$; then $\zeta_1^t, \zeta_2^t, \ldots, \zeta_s^t$ are also such a basis.

__Proof.__ Let $\Sigma a_i \zeta_i^t = 0$ with a_i in $GF(q^m)$. Apply Lemma 5.7 with $c_1 = c_2 = \cdots = c_s = 0$, this shows that $a_1 = a_2 = \cdots = a_s = 0$. Hence $\zeta_1^t, \zeta_2^t, \ldots, \zeta_s^t$ are linearly independent over $GF(q^m)$ and form a basis for $GF(q^{N+1})$.

__Lemma 5.9.__ Suppose (5.13) holds with $(t,v) = 1$. Let ω be an element of $GF(q^{N+1})$ and suppose $N + 1 > m$. Then $(1 + \omega)^t = a_1 + a_2 \omega^t$ for some elements a_1, a_2 of Q^*.

__Proof.__ First assume that ω is not an element of $GF(q^m)$. Then there exists a basis for $GF(q^{N+1})$ over $GF(q^m)$ which contains $\zeta_1 = 1$, $\zeta_2 = \omega$. Let $c_1 = c_2 = 1$, $c_3 = \cdots = c_s = 0$; so $1 + \omega = \Sigma c_i \zeta_i$. Moreover $\zeta_1^t, \zeta_2^t, \ldots, \zeta_s^t$ is also a basis (Lemma 5.8) so

$$(1 + \omega)^t = \Sigma a_i \zeta_i^t$$

with a_i in $GF(q^m)$. By Lemma 5.7, $a_3 = a_4 = \cdots = a_s = 0$, and a_1, a_2 belong to Q^*.

Now suppose ω is in $GF(q^m)$. Let ζ be an element of $GF(q^{N+1})$ such that

$\tilde{L}(\zeta) = \omega$ and ζ does not belong to $GF(q^m)$. [Such a ζ exists since $\tilde{L}(1) = 1$ and so $\tilde{L}(\omega) = \omega$; thus $\zeta = \omega + \varphi$ where φ does not belong to $GF(q^m)$ and $\tilde{L}(\varphi) = 0$. Now $N + 1 > m$ and $\tilde{L}(1) = 1$ guarantee the existence of such an element φ.] Now $\tilde{L}(1 + \zeta) = 1 + \omega$ and so (1) $\tilde{L}(\eta(1 + \zeta)^t)$ belongs to $(1 + \omega)^t$ Q^* by Lemma 5.6. On the other hand, by the first part of this proof $(1 + \zeta)^t = b + c\zeta^t$ for some elements b, c of Q^*. Thus

$$\tilde{L}(\eta(1 + \zeta)^t) = \tilde{L}(\eta(b + c\zeta^t)) = b\,\tilde{L}(\eta) + c\,\tilde{L}(\eta\,\zeta^t)$$

and $\tilde{L}(\eta) = \tilde{L}(\eta\, 1^t)$ belongs to Q^* while $\tilde{L}(\eta\,\zeta^t)$ belongs to $\omega^t Q^*$ since $\tilde{L}(\zeta) = \omega$. Thus (2) $\tilde{L}(\eta(1 + \zeta)^t)$ is of the form $a_1 + a_2\,\omega^t$ with a_1, a_2 in Q^*. Combining this with (1) above completes the proof.

Lemma 5.9 is the first major step in the Gordon, Mills, Welch proof of Theorem 5.12 below. To complete the proof some results about the implications of the condition

$$(1 + \omega)^t = a_1 + a_2\,\omega^t \tag{5.17}$$

are required. If (5.17) holds for all ω in $GF(q^{N+1})$, then it follows that for every pair ρ_1, ρ_2 of elements of $GF(q^{N+1})$, there exist elements b_1, b_2 of Q^*, such that $(\rho_1 + \rho_2)^t = b_1\,\rho_1^t + b_2\,\rho_2^t$. By induction, given any $\rho_1, \rho_2, \ldots, \rho_u$ in $GF(q^{N+1})$ there exist $b_1, b_2, \ldots, b_u$ in Q^* such that

$$(\rho_1 + \rho_2 + \cdots + \rho_u)^t = b_1\rho_1^t + b_2\rho_2^t + \cdots + b_u\rho_u^t. \tag{5.18}$$

Write (5.17) in the form

$$(1 + \omega)^t = r_\omega(1 + s_\omega\,\omega^t)$$

with r_ω, s_ω in Q^*. Since $(t,v) = 1$, it follows that if ω does not belong to Q then neither does ω^t and hence r_ω, s_ω are uniquely determined.

Lemma 5.10. Let $N \geq 2$ and, for every ω in $GF(q^{N+1})$, let there exist elements a_1, a_2 of Q^* such that (5.17) holds. Then

(i) if ω^t, τ^t, ζ^t are linearly independent over Q, then $s_{\zeta/\omega} = s_{\zeta/\tau}\, s_{\tau/\omega}$.

(ii) if α is a primitive root of $GF(q^{N+1})$ and if $s_\alpha = 1$, then for all ω which are not elements of Q, $s_\omega = 1$ and $(1 + \omega)^t = 1 + \omega^t$.

Proof. For uniquely determined b_1, b_2, b_3, c_1, c_2, c_3 of Q^*

$$b_1\omega^t + b_2\tau^t + b_3\zeta^t = (\omega + \tau + \zeta)^t = c_1(\omega + \tau)^t + c_2\zeta^t$$

$$= c_1\omega^t(1 + \tau/\omega)^t + c_2\zeta^t = c_3(\omega^t + s_{\tau/\omega}\,\tau^t) + c_2\zeta^t .$$

So $s_{\tau/\omega} = b_2/b_1$ and by symmetry $s_{\zeta/\tau} = b_3/b_2$, $s_{\zeta/\omega} = b_3/b_1$ which establishes part (i) of the lemma.

Consider the $s_\omega = 1$ assertion of (ii). Since $(t,v) = 1$ it follows that α^t is not contained in any proper subfield of $GF(q^{N+1})$. Hence α^t has degree $N + 1$ over Q. Hence 1, α^t and α^{2t} are linearly independent over Q (as $N + 1 \geq 3$). Put $\omega = \alpha^u$, $1 \leq u < q^{N+1} - 1$ and induct on u. By part (i) it follows that

$$s_{\alpha^2} = s_\alpha\, s_\alpha = 1$$

so $s_\omega = 1$ when $u = 1$ or 2. Let $u \geq 3$ and suppose that $s_\omega = 1$ for all positive integers less than u. Since ω is not an element of Q neither is ω^t, as $(t,v) = 1$. So 1 and ω^t are linearly independent over Q. Now the elements 1, α^t, α^{ut} or the elements 1, α^{2t}, α^{ut}, are linearly independent over Q; for if both these element sets are dependent then so is the set 1, α^t, α^{2t} (contradiction). Let 1, α^{jt}, α^{ut} be independent ($j = 1$ or 2), then by part (i) and the induction hypothesis

$$s_\omega = s_{\alpha^u} = s_{\alpha^{u-j}}\, s_{\alpha^j} = 1$$

and so the first assertion of (ii) has been established.

Since $N + 1 \geq 3$, there is a ζ such that $1, \omega^t, \zeta^t$ are linearly independent over Q. Then, for suitable c_1, c_2, c_3 in Q^*.

$$(1 + \omega + \zeta)^t = c_1 + c_2\omega^t + c_3\zeta^t$$

with c_1, c_2, c_3 uniquely determined. Since $s_\omega = 1$, $(1 + \omega)^t = r_\omega(1 + \omega^t)$, and by the linear independence of $1, \omega^t, \zeta^t$ it follows that $(1 + \omega)^t/\zeta^t$ is not in Q. So $(1 + \omega)/\zeta$ is not in Q. Thus

$$(1 + \omega + \zeta)^t = a(1 + \omega)^t + a\zeta^t = ar_\omega + ar_\omega\omega^t + a\zeta^t$$

where $a = r_{(1+\omega)/\zeta}$. Hence $c_1 = c_2$. By the same process it follows that $c_1 = c_3$ and from this that $r_\omega = 1$. So $(1 + \omega)^t = 1 + \omega^t$ as was to be shown.

<u>Theorem 5.11</u>. Let $N + 1 \geq 3$, let q be a power of the prime p, let $v = (q^{N+1} - 1)/(q - 1)$ and let t be an integer relatively prime to v. Suppose that for every ω in $GF(q^{N+1})$ there exist non-zero elements a_1, a_2 in $GF(q)$ such that $(1 + \omega)^t = a_1 + a_2\omega^t$. Then t is congruent to a power of p modulo v.

<u>Proof</u>. Without loss of generality $0 < t < v$ and $(1 + \alpha)^t = r_\alpha(1 + s_\alpha\, \alpha^t)$, where α is a fixed primitive root of $GF(q^{N+1})$. Since s_α is in Q^* it follows that $s_\alpha = \alpha^{vc}$ for some c, $0 \leq c < q - 1$. Put $t' = t + vc$. Then $0 < t' < q^{N+1} - 1$ and $(t', v) = 1$. Furthermore, for any ω in $GF(q^{N+1})$ there exist r'_ω and s'_ω in Q^* such that

$$(1 + \omega)^{t'} = r'_\omega(1 + s'_\omega\, \omega^{t'}) .$$

Note that $s'_\alpha = 1$. Thus by part (ii) of Lemma 5.10

$$(1 + \omega)^{t'} = 1 + \omega^{t'} \tag{5.19}$$

for all ω of $GF(q^{N+1})$ which are not elements of Q. Suppose t' is not a power of p. Then (5.19) becomes a polynomial equation of degree at most $t' - 1$, with at least $q^{N+1} - q$ roots. Therefore $t' > q^{N+1} - q$. Let $u = q^{N+1} - 1 - t'$. Multiplying (5.19) by $\omega^u(1 + \omega)^u$ yields

$$\omega^u = (1 + \omega)^u \omega^u + (1 + \omega)^u$$

for all ω of $GF(q^{N+1})$ which do not belong to Q. Hence $2u \geq q^{N+1} - q$. So

$$q^{N+1} - 1 = t' + u > \frac{3}{2} (q^{N+1} - q)$$

or $q^{N+1} < 3q - 2$ which is impossible since $N + 1 \geq 3$, $q \geq 2$. Thus t' is a power of p and t is congruent to a power of p modulo v, which completes the proof. Finally, the result promised above can be established.

<u>Theorem 5.12</u>. (Gordon, Mills and Welch) Let q be a power of a prime p and let $N + 1$ be a positive integer such that $N + 1 > m \geq 2$, where m divides $N + 1$. Let $v, k, \lambda, w, \ell, \mu$ be given by (5.5) and (5.10), let $\xi = v/w$, and let $\Omega(x)$ be the polynomial given by (5.7). To any w,ℓ,μ-difference set B with Hall polynomial $\psi(y)$, there corresponds a v,k,λ-difference set $\bar{B}$ with Hall polynomial $\theta(x) = \Omega(x)\,\theta(x^{\xi})$. If B and C are w,ℓ,μ-difference sets then $\bar{B}$ and $\bar{C}$ are equivalent if and only if B is a cyclic shift of C.

<u>Proof</u>. Start with B and C and construct the Hall polynomials $\theta_b(x)$, $\theta_c(x)$ of $\bar{B}$, $\bar{C}$ respectively. If $\bar{B}$ is equivalent to $\bar{C}$ then there exist integers a, t with $(t,v) = 1$ such that

$$\theta_b(x) \equiv x^a \theta_c(x^t) \qquad (\text{mod } x^v - 1) .$$

Hence, since $N + 1 \geq 3$, Lemma 5.4, Lemma 5.9 and Theorem 5.11 establish that t is congruent to a power of p modulo v. Now every power of p is a multiplier of B and C (apply Hall's Theorem 3.1 above to their complements). So

$$\psi_c(y^t) \equiv y^u \psi_c(y) \qquad \pmod{y^w - 1}$$

for some integer u. Thus by Lemma 5.4

$$\psi_b(y) \equiv y^{u+s} \psi_c(y) \qquad \pmod{y^w - 1}$$

and B is a cyclic shift of C as promised.

In the course of the above proof it was concluded that, for $N + 1 \geq 3$,

$$\theta_b(x) \equiv x^a \theta_c(x^t) \qquad \pmod{x^v - 1}$$

only happens when t is congruent to a power of p modulo v. If $B = C$ this same observation shows that only powers of p may be multipliers of the difference sets $\overline{B}$. With $m = 1$ this shows that only the powers of p may be multipliers of the $D(L,\alpha)$ difference sets. As mentioned earlier, Theorem 3.1 shows that the powers of p are always multipliers of these difference sets, so:

Theorem 5.13. (Gordon, Mills and Welch) If D is a non-trivial Singer difference set or if D is any non-trivial difference set derived from congruence 5.6, then the multipliers of D are precisely the powers of p modulo v.

Another by-product of Theorem 5.12 is the existence of parameters v, k, λ for which at least $j > 0$ inequivalent difference sets exist. For, let $m \geq 3$, and let J denote the number of inequivalent w, ℓ, μ-difference sets with w, ℓ, μ given by (5.10). Since -1 is never a multiplier of a non-trivial difference set (see Theorem 3.3) there are at least $2J$ possible w, ℓ, μ-difference sets, none of which is a cyclic shift of any other. Hence there are at least $2J$ inequivalent v, k, λ-difference sets with v, k, λ given by (5.5) or (5.1). In particular

Theorem 5.14. (Gordon, Mills, and Welch) Let q be any prime power and let m, M be positive integers with $m \geq 3$. Let $N + 1 = Mm$ and let M be the product of r prime numbers, not necessarily distinct. Then there exist at least 2^r

inequivalent difference sets with parameters (5.5) and thus (by taking complements) with parameters

$$v = \frac{q^{N+1} - 1}{q - 1}, \quad k = \frac{q^{N} - 1}{q - 1}, \quad \lambda = \frac{q^{N-1} - 1}{q - 1}.$$

B. N^{th} Power Residue Difference Sets and Cyclotomy

A difference set which is composed of all the Nth powers modulo some prime v, or of the Nth powers and zero, is called an Nth power residue difference set. [Those containing zero are called modified N^{th} power residue difference sets, when there is some reason for distinguishing the two types.] Attention may be restricted to divisors N of $v - 1$, since if $(N, v - 1) = d \geq 1$ the Nth power residues and the dth power residues coincide. For primes $v = Nf + 1$, the fact that the f distinct N^{th} power residues $r_1, \ldots, r_f$ form a subgroup of the group of non-zero residues modulo v will be used extensively [Nagell (1951) is a convenient reference for the number theory required in this section]. The best known N^{th} power residue difference sets are the quadratic residue sets of Paley (1933):

Theorem 5.15. When $v = 4t - 1$ is a prime, the quadratic residues modulo v form a difference set with parameters $v, k, \lambda, n = 4t - 1, 2t - 1, t - 1, t$.

Proof. Recall from number theory that exactly one of $c, -c$ is a quadratic residue for a prime of this form, $1 \leq c \leq v - 1$. Thus one of $r_1 - r_2, r_2 - r_1$ is a quadratic residue r, say $r_1 - r_2 \equiv r$. If s is any quadratic residue of v, then sr_1, sr_2, sr are quadratic residues also. Thus every equation $r_i - r_j \equiv r$ corresponds with an equation $sr_i - sr_j \equiv sr$ and vice versa. Hence every quadratic residue of v is represented equally often as a difference of quadratic residues. Reversing these congruences (e.g., $r_2 - r_1 \equiv -r$) yields every quadratic non-residue with the same number of representations also. Hence this is a difference set with $v, k = 4t - 1, 2t - 1$; thus $\lambda = t - 1$.

The set $\{1,2,4\}$ modulo 7 is the first non-trivial quadratic residue difference set. Note that the parameters imply that every quadratic residue set is a Hadamard difference set (see section IV.B. for a discussion of this special type).

Another class of residue difference sets was discovered by Chowla (1944):

Theorem 5.16. The biquadratic residues of primes $v = 4x^2 + 1$, x odd, form a difference set with parameters $v,k,\lambda = 4x^2 + 1,\ x^2,\ (x^2 - 1)/4$.

The first non-trivial biquadratic residue difference set is for $x = 3$, here $v,k,\lambda = 37,9,2$ and $D = \{1,7,9,10,12,16,26,33,34\}$.

Other N^{th} power residue difference sets have been discovered and a general theory for them has been developed by Emma Lehmer (1953). In order to explain her results a little of the theory of cyclotomy will have to be introduced. No more cyclotomy than is necessary for an understanding of the difference set results of sections V.B., V.C., and V.D. is developed here. [A complete introduction to cyclotomy from the classical point of view is given ab initio in Dickson (1935 ABC); T. Storer's booklet "Cyclotomy and Difference Sets" (1967 A) gives a different development also ab initio and discusses most of the difference set results of these sections.]

Let $v = Nf + 1$ be an odd prime and let g be a fixed primitive root of v. An integer R is said to belong to the index class ℓ with respect to g if there exists an integer x such that $R \equiv g^{Nx+\ell} \pmod v$. Thus, the index class ℓ consists of f distinct numbers $g^{\ell}, g^{N+\ell}, \ldots, g^{N(f-1)+\ell}$ modulo v. The cyclotomic number $(\ell,m)_N$ counts the number of times $R + 1$ belongs to index class m when R belongs to index class ℓ. That is, $(\ell,m)_N$ is the number of solutions x, y of the congruence

$$g^{Nx+\ell} + 1 \equiv g^{Ny+m} \pmod v \qquad (5.20)$$

where the integers x, y are chosen from $0,1,\ldots,f-1$. This congruence shows that there are at most N^2 distinct cyclotomic numbers of order N and that these numbers depend not only on v, N, ℓ, m but also on which of the $\varphi(v-1)$

primitive roots g of v is chosen.

The following elementary cyclotomic facts are all that is needed for an understanding of the results of this section (note that when f is odd, N is necessarily even):

$$(\ell,m)_N = (\ell',m')_N \quad \text{when} \quad \ell \equiv \ell' \quad \text{and} \quad m \equiv m' \qquad (\text{mod } N) \tag{5.21}$$

$$(\ell,m)_N = (N-\ell,\ m-\ell)_N = \begin{cases} (m,\ell)_N & f \text{ even} \\ (m+N/2,\ \ell+N/2)_N & f \text{ odd} \end{cases} \tag{5.22}$$

$$\sum_{m=0}^{N-1} (\ell,m)_N = f - n_\ell, \quad \text{where } n_\ell = \begin{cases} 1 & \ell \equiv 0 \pmod N \quad f \text{ even} \\ 1 & \ell \equiv N/2 \pmod N \quad f \text{ odd} \\ 0 & \text{otherwise} \end{cases} \tag{5.23}$$

$$(\ell,m)'_N = (s\ell, sm)_N \tag{5.24}$$

where $(\ell,m)'_N$ is based on the primitive root $g' \equiv g^s$ modulo v; necessarily then s is prime to $v-1$.

Proof. Equation (5.21), of course, is an immediate consequence of the definition. The first part of (5.22) follows from the definition after congruence 5.20 is multiplied through by the inverse of its first term [i.e., by $g^{N(f-x)-\ell}$]. Similarly, the second part of (5.22) follows after congruence 5.20 is multiplied through by -1, that is by

$$-1 \equiv g^{(Nf)/2} = \begin{cases} g^{N(f/2)} & f \text{ even} \\ g^{N(f-1)/2+N/2} & f \text{ odd}. \end{cases} \tag{5.25}$$

The sum in equation (5.23) is simply the number of successors of members of index class ℓ which belong to any index class at all. Since -1 is the only element whose successor does not belong to an index class, equation (5.25) implies the correctness of (5.23). Equation (5.24) follows from the definition, since s prime to $v - 1$ implies the sx ranges with x over a complete set of residues modulo f.

In terms of these cyclotomic numbers $(\ell,m)_N$ it is possible to give necessary and sufficient conditions for the existence of Nth power residue difference sets; in fact:

Theorem 5.17. (Lehmer, 1953) Necessary and sufficient conditions, that the Nth power residues of a prime $v = Nf + 1$ form a difference set, are that N is even, f is odd and that

$$(\ell,0)_N = (f - 1)/N \qquad \text{for} \qquad \ell = 0,1,\ldots,\tfrac{1}{2}N - 1. \tag{5.26}$$

The parameters of such difference sets are $v,k,\lambda = v,f, (f - 1)/N$. Necessary and sufficient conditions, that the Nth power residues and zero for a prime $v = Nf + 1$ form a difference set, are that N be even, f odd and that

$$1 + (0,0)_N = (\ell,0)_N = (f + 1)/N \qquad \text{for} \qquad \ell = 1,2,\ldots, \tfrac{1}{2}N - 1. \tag{5.27}$$

The parameters of such difference sets are $v,k,\lambda = v,\ f + 1,\ (f + 1)/N$.

[Note that equations (5.22) and (5.24) show that these existence conditions (5.26) and (5.27) are independent of the primitive root g, even though the individual cyclotomic numbers $(\ell,0)_N$ are not.]

Proof. If the Nth power residues $r_1,\ldots,r_f$ are a difference set modulo v, then there are exactly λ solutions r_i, r_j to the congruence

$$r_i - r_j \equiv \gamma \qquad \pmod{v} \tag{5.28}$$

and hence the congruence

$$r_i r_j^{-1} \equiv \gamma r_j^{-1} + 1 \qquad (\mathrm{mod}\ v) \qquad (5.29)$$

for all $\gamma \neq 0$ modulo v. But for γ in index class c this implies that $(c,0)_N = \lambda$; thus $(i,0)_N = \lambda$ for $0 \leq i \leq N - 1$ and in particular (5.26) holds as $\lambda = (f - 1)/N$ follows from $v,k = v,f$. If f is even, then by (5.25), -1 is an Nth power residue, hence a multiplier, which contradicts Theorem 3.3. Thus f is odd and as Nf is necessarily even, so is N. Conversely, since N is even and f is odd, equation (5.22) provides $(i,0)_N = (i + \frac{N}{2},\ 0)_N = (f - 1)/N$ for all i. Hence (5.29) and (5.28) have exactly $(f - 1)/N$ solutions for all γ. That is, the Nth powers modulo v form a difference set with parameters $v,k,\lambda = v, f, (f - 1)/N$.

When 0 is added to the set of Nth power residues the only effect is that differences

$$r_i - 0 = r_i \quad \text{and} \quad 0 - r_i = -r_i$$

have to be counted also. As before, f even implies that -1 is a multiplier. So, f is odd (hence N is even) and -1 belongs to index class $N/2$; thus each r_i and $-r_i$ is represented once more than the numbers $(0,0)_N$ and $(N/2, 0)_N$ indicate. In particular, equation (5.27) holds, since $\lambda = (f+1)/N$ follows from $v,k = v, f + 1$. On the other hand, assume that N is even, f is odd and that equation (5.27) holds. Since $(i + \frac{N}{2}, 0)_N = (i,0)_N$ here [by equation (5.22)] it follows, as before, that the Nth power residues and zero form a difference set with $v,k = v, f + 1$; hence $\lambda = (f + 1)/N$.

Theorem 5.17 has been applied for several values of N, i.e., those for which the cyclotomic numbers have been computed, with the following results:

<u>Theorem 5.18</u>. Difference sets exist which consist of (i) [$N = 2$, Paley (1933)] the quadratic residues of primes $v = 4t - 1$; e.g., the $v,k,\lambda = 7,3,1$ set $\{1,2,4\}$, (ii) [$N = 4$, Chowla (1944)] the biquadratic residues of primes $v = 4x^2 + 1$, x odd; e.g., the $v,k,\lambda = 39,9,2$ set $\{1,7,9,10,12,16,26,33,34\}$,

(iii) [N = 4, Lehmer (1953) attributes these to M. Hall, Jr.] the biquadratic residues and zero for primes $v = 4x^2 + 9$, x odd; e.g., the $v,k,\lambda = 13,4,1$ set {0,1,3,9}, (iv) [N = 8, Lehmer (1953)] the octic residues of primes $v = 8a^2 + 1 = 64b^2 + 9$, $k = a^2$, $\lambda = b^2$ with a, b odd; e.g., the $v,k,\lambda = 73,9,1$ set {1,2,4,8,16,32,37,55,64}, the next such prime v is 140,411,704,393, (v) [N = 8, Lehmer (1953)] the octic residues and zero for primes $v = 8a^2 + 49 = 64b^2 + 441$, $k = a^2 + 7$, $\lambda = b^2 + 7$, a odd, b even; e.g., the $v,k,\lambda = 26041,3256,407$ set, there are no more such primes $v < 34,352,398,777$.

On the other hand (by these methods) it can be shown that <u>no</u> modified quadratic residue difference sets exist and further that <u>no</u> non-trivial residue or modified residue difference sets exist when (vi) N = 6, Lehmer (1953), (vii) N = 10, Whiteman (1960 B), (viii) N = 12, Whiteman (1960A), (ix) N = 14, Muskat (1966), (x) N = 16, Whiteman (1957) for the case when 2 is an octic residue of v. The case where 2 is <u>not</u> an octic residue is open; however, it is known that if the residues form a difference set here then $v \geq 1,336,337$.
(xi) N = 18, Baumert and Fredricksen (1967), (xii) N = 20, Muskat and Whiteman (submitted for publication) for the case when 5 is a biquadratic residue of v. The case where 5 is <u>not</u> a biquadratic residue remains unsolved.

The calculations involved in determining the cyclotomic numbers of these orders are far too extensive to present here, the numbers themselves may be found for N = 2 - 6 in Dickson (1935 A), N = 8 in Lehmer (1955 B), and in the papers cited above for N = 10,12,14,16,18,20.

By way of example, though, if N = 2 and f odd, equation (5.22) yields $(0,0)_2 = (1,1)_2 = (1,0)_2$ thus (5.23) provides two linear relations, which may be solved for $(0,0)_2 = (1,1)_2 = (1,0)_2 = (f - 1)/2$ and $(0,1)_2 = (f + 1)/2$. So (5.26) of Theorem 5.17 provides no restrictions, hence the quadratic residues of every prime $v = 2f + 1$, f odd, form a difference set [i.e., an alternate proof of Theorem 5.15 has been given], and (5.27) prohibits the existence of modified quadratic residue difference sets. In the case N = 4, f odd, the computation of the cyclotomic numbers is more difficult; however, they are

$$(0,0)_4 = (2,2)_4 = (2,0)_4 = (v - 7 + 2y)/16$$

$$(0,1)_4 = (1,3)_4 = (3,2)_4 = (v + 1 + 2y - 8x)/16$$

$$(1,2)_4 = (0,3)_4 = (3,1)_4 = (v + 1 + 2y + 8x)/16$$

$$(0,2)_4 = (v + 1 - 6y)/16$$

and the rest are $(v - 3 - 2y)/16$, where $v = y^2 + 4x^2$ and $y \equiv 1$ modulo 4. Thus (5.26) of Theorem 5.17 requires $v - 7 + 2y = v - 3 - 2y = v - 5$ or $y = 1$. [That is, a proof of Theorem 5.16 has been given.] The existence of modified biquadratic residue difference sets requires, by (5.27), that $v + 2y + 9 = v - 3 - 2y = v + 3$ or $y = -3$, thus $v = 9 + 4x^2$. Since $v = 4f + 1$, f odd, let $f = 2j + 1$, then $v = 8j + 5 = 9 + 4x^2$. Thus $8j - 4x^2 = 4$, $2j - x^2 = 1$; hence x is odd, which establishes part (iii) of Theorem 5.18.

A completely analoguous theory of cyclotomy can be developed for prime powers $v = p^i = Nf + 1$ [see, for example, Hall (1965) or Storer (1967 A)]. This results (for $N = 2,4,6,8$) in the same conditions for N^{th} power and modified N^{th} power residue difference sets as before. However (for $i > 1$) these difference sets are not cyclic and, in fact, for $N = 4,8$ the quadratic conditions imposed by Theorem 5.18 are only satisfied for $i = 1$, i.e., v is prime. [For the biquadratic residues, this was established by A. U. Lebesque (1850) and recently rediscovered by Hall (1965). Hall's paper also contains a proof for the biquadratic residues plus 0. Storer (1967 A, Theorem 20) states this result without proof for $N = 8$.]

The multiplier group for N^{th} power residue difference sets has been determined:

Theorem 5.19. (Lehmer, 1953) The Nth power residues themselves are the only multipliers of a non-trivial N^{th} power residue difference set.

Proof. Clearly the Nth power residues are multipliers. Suppose t was another multiplier, then $tD = D + s$ modulo v, where t belongs to an index class j, with $j \not\equiv 0 \pmod N$. In fact, the index class j consists of $tr_1, \ldots, tr_f$; so $s \not\equiv 0$ modulo v. Let s belong to index class i. Then, if 0 is not in D, $r_u + s \equiv tr_y \pmod v$ implies

$$r_u s^{-1} + 1 \equiv tr_y s^{-1} \qquad \pmod v$$

has exactly f solutions, i.e., the cyclotomic number $(N - i, j - i)_N = (i,j)_N = f$. But this implies, by (5.23), that $(i,0)_N = 0$, which contradicts the existence condition (5.26) when D is non-trivial.

If 0 belongs to D, then $tD = D + s$ (with $s \not\equiv 0$) implies $0 - s$ belongs to D. Thus, by (5.25), s belongs to index class $N/2$. On the other hand, $tr_u - s \equiv 0$ for some u; thus t and -1 belong to the same index class. Hence -1 is a multiplier also, contradiction.

As mentioned in sections I.E. and III.A. above, every divisor of $n = k - \lambda$ is a multiplier for all known cyclic difference sets. This is not readily apparent for the residue difference sets of Theorem 5.18, but was established by Emma Lehmer (1955 A). First consider the quadratic residue difference sets for prime $v = 4t - 1$; for these it is necessary to use the law of quadratic reciprocity [see, for example, Nagell (1951)]. As usual, the Legendre symbol $(p/q) = \pm 1$ will be used to indicate whether ($= +1$) or not ($= -1$) p is a quadratic residue of the odd prime q.

Theorem 5.20. (Lehmer, 1955 A) All divisors of n $(= k - \lambda)$ are multipliers for the quadratic residue difference sets for primes $v = 4t - 1$.

Proof. By Theorem 5.19, it is sufficient to show that all divisors of n are quadratic residues modulo v, which, of course, follows if the prime divisors of n are quadratic residues. Now $n = (v + 1)/4$ and 2 divides n only if $v \equiv -1 \pmod 8$, but then 2 is a quadratic residue by the reciprocity law. Let q be an odd prime divisor of n (and thus of $v + 1 = qm$). By quadratic reciprocity

$$\left(\frac{q}{v}\right) = \left(\frac{v}{q}\right)(-1)^{(q-1)(v-1)/4} = \left(\frac{qm-1}{q}\right)\left(\frac{-1}{q}\right)^{(v-1)/2} = \left(\frac{-1}{q}\right)^2$$

i.e., q is a quadratic residue of v.

More generally there is:

Theorem 5.21. (Lehmer, 1955 A) All divisors of n are multipliers for the cyclic N^{th} power residue difference sets of Theorem 5.18.

C. More Cyclotomic Difference Sets

Denote each index class i by C_i, then if $v = 31$ and $N = 6$ these classes are:

C_0:	1,	2,	4,	8,	16
C_1:	3,	6,	12,	17,	24
C_2:	5,	9,	10,	18,	20
C_3:	15,	23,	27,	29,	30
C_4:	7,	14,	19,	25,	28
C_5:	11,	13,	21,	22,	26

Letting $C_0 + C_1 + C_3$ denote the union of these classes, it can be verified that $C_0 + C_1 + C_3$ forms a difference set with parameters $v,k,\lambda = 31,15,7$. This is a special case of Theorem 5.23 given below. Note that if a union of index classes of the Nth power residues of some odd prime $v = Nf + 1$ form a difference set, then it is immediate that all the Nth power residues are multipliers. Since -1 is an Nth power residue if f is even (in fact, all solutions of $x^f \equiv 1$ modulo v are N^{th} power residues) such a union of index classes can only form a difference set for f odd (see Theorem 3.3 above). Thus:

Lemma 5.22. A union of index classes (with or without 0 added) of the Nth power residues of some odd prime $v = Nf + 1$ can form a difference set only if f is odd and N is even.

If the union of index classes $C_i + C_j + \cdots + C_m$ forms a difference set then so does $C_{i+s} + C_{j+s} + \cdots + C_{m+s}$ for any s; this second difference set can be obtained from the first by multiplying by any element of C_s and thus is equivalent to the first. (This fact considerably shortens any exhaustive examination of index class unions.)

Let $N = 2$, then with $k < v/2$, as usual, only C_0 need be considered and this leads to the quadratic residue difference sets of paragraph V.B. For $N = 4$ (with $k < v/2$) only $C_0, C_0 + 0$, $C_0 + C_2$ and $C_0 + C_1$ need be considered. The first three of these correspond to the biquadratic residues, the biquadratic residues plus 0 and the quadratic residues; which of course constitute difference sets under the conditions given in Theorem 5.18 above. The cyclotomic numbers discussed in section V.B. will be used to rule out the remaining possibility.

In general, congruence 5.20 shows that the cyclotomic number $(\ell,m)_N$ is the number of solutions of

$$\alpha - \beta \equiv 1 \pmod{v}$$

for α in C_ℓ and β in C_m. Thus $(\ell - s, m - s)_N$ is the number of solutions of

$$g^{Nx+\ell-s} - g^{Ny+m-s} \equiv 1 \pmod{v}$$

and therefore of

$$g^{Nx+\ell} - g^{Ny+m} \equiv d \pmod{v}$$

for fixed d in C_s. Thus, the number of solutions of

$$\alpha - \beta \equiv d \pmod{v} \tag{5.30}$$

for fixed d in C_s with α in C_ℓ, β in C_m is $(\ell - s, m - s)_N$. So, for index classes $C_{z_1}, \ldots, C_{z_h}$, the number of solutions of congruence 5.30 for fixed d in C_s is

$$J_s = \sum_{i=1}^{h} \sum_{j=1}^{h} (z_i - s, z_j - s)_N . \qquad (5.31)$$

Thus a difference set corresponds to $C_{z_1} + \cdots + C_{z_h}$ if and only if $J_0 = J_1 = \cdots = J_{N-1} = \lambda$. Since f is odd, $J_i = J_{i+N/2}$ (by equation (5.22)) so only $J_0 = \cdots = J_{(N/2)-1} = \lambda$ need be checked.

Now consider $C_0 + C_1$ for $N = 4$, then

$$J_0 = (0,0)_4 + (1,0)_4 + (0,1)_4 + (1,1)_4 = (4v - 12 - 8x)/16$$

$$J_1 = (-1,-1)_4 + (0,-1)_4 + (-1,0)_4 + (0,0)_4 = (4v - 12 + 8x)/16$$

as the values of $(\ell,m)_4$ given in section V.B. show. Thus $J_0 = J_1$ only if $x = 0$; but $v\ (= y^2 + 4x^2)$ is y^2 then, so there is no prime v for which $C_0 + C_1$ form a difference set with $N = 4$.

When 0 is to be added to the index class union, differences of the type $d - 0$ and $0 - d$ must be counted also. Since f is odd, -1 belongs to class $N/2$ (congruence 5.25 above); thus J_s should be increased by 1 for each index class C_i in the union for which $i \equiv s$ modulo $N/2$.

Hall (1956) applied this technique to the case $N = 6$ and established:

Theorem 5.23. A set of residues forming a non-trivial difference set modulo a prime $v = 6f + 1$ which includes the sextic residues as multipliers may consist of (1) the qudratic residues for $v \equiv 3$ modulo 4 or (2) the index classes C_0, C_1 and C_3 for an appropriate choice of primitive root g of v whenever v is representable as $4x^2 + 27$. The only possibilities are equivalent to one of the above. (The appropriate primitive root g puts the residue 3 in C_1.)

For $N = 10$ the same technique was applied by Hayashi (1965) yielding:

Theorem 5.24. A set of residues forming a non-trivial difference set modulo a prime $v = 10f + 1$ which includes the 10th power residues as multipliers may consist of (1) the quadratic residues for $v \equiv 3$ modulo 4 or (2) the index classes C_0 and C_1 for the primitive root $g = 11$ when $v = 31$.

Yamamoto (1967) developed a test for deciding whether a union of s such N^{th} power residue index classes (with or without 0 added) forms a difference set. His test has the advantage that it does not require the prior determination of the cyclotomic numbers of order N. The initial form of his result is:

Theorem 5.25. Let $v = Nf + 1$ be prime with N even and f odd. Let B be a subset of the set of fth power residues of v, containing exactly s such residues. Let E be the set of all residues $a \pmod{v}$ such that a^f is in B. Let $d = 0$ or 1 and define $D = E$ of $D = E$ with 0 added according as $d = 0$ or 1. Then the set D is a difference set if and only if

$$s(sf + 2d - 1) \equiv 0 \pmod{N} \tag{5.32}$$

$$\sum_{i=0}^{j} (-1)^i \binom{jf}{if} K_i K_{j-1} \equiv 2dNK_j \pmod{v} \tag{5.33}$$

for $j = 2,4,\ldots,N - 2$, where K_r is the sum of the rth powers of the elements of B.

After modifying congruence 5.33, Yamamoto applied this test for $N = 4,6,8,10,12$ with the result:

Theorem 5.26. A set of residues forming a non-trivial difference set modulo a prime $v = Nf + 1$ for $N = 4,6,8,10$ or 12, which includes the Nth power residues as multipliers, may consist of the quadratic residues, the biquadratic residues (with or without 0), the octic residues (with or without 0), the Hall sets for $v = 4x^2 + 27$ or the special 31, 6, 1 set of Theorem 5.24.

In a further attack on the problem, Yamamoto (1969) develops a slightly different condition, and, by way of example, uses it to show that there exist no non-trivial residue or modified residue difference sets for $N = 6,10,14$. [I.e., he reestablishes parts (vi), (vii), (ix) of Theorem 5.18, section V.B. above.]

In addition to the above facts it is known (Baumert and Fredricksen, 1967) that there are 6 index class unions of the 18th power residues for $v = 127$ which lead to inequivalent $v,k,\lambda = 127,63,31$ difference sets. For $N = 6$ Hall (1965) developed the requisite prime power cyclotomy and generalized his $v = 4x^2 + 27$ difference sets to prime powers v. However, as is shown there [proof due to W. H. Mills], the only prime powers $4x^2 + 27$ are in fact primes. These difference sets, $v = 4x^2 + 27$ of Hall, were examined from the point of view of the multiplier problem and it was found that indeed every divisor of n was a multiplier for these sets. [Emma Lehmer (1955 A) gives a proof of this fact and attributes an earlier proof by means of cubic reciprocity to Hall.]

D. Generalized Cyclotomy and Difference Sets

In 1958 Stanton and Sprott (1958) published a generalization of the following result:

Theorem 5.27. Let g be a primitive root of both p and $p + 2$, where p and $p + 2$ are both prime. Then the numbers

$$1, g, g^2, \ldots, g^{(p^2-3)/2}; \quad 0,\ p+2,\ 2(p+2), \ldots,\ (p-1)(p+2)$$

form a difference set with parameters $v,k,\lambda = p(p+2),\ (v-1)/2,\ (v-3)/4$, i.e., a Hadamard difference set.

These difference sets (the so-called twin prime sets) were in fact already known, although in slightly different guise. They had been independently discovered not only by Stanton and Sprott but also by Kesava Menon (1962), Brauer (1953), Chowla (1945), perhaps first by Gruner (1939) and probably a few others, as they seem to belong to that special class of mathematical objects which are prone to independent rediscovery.

Motivated by Theorem 5.27, Whiteman (1962) investigated the problem of when the $k = d + p$ numbers

$$1, g, g^2, \ldots, g^{d-1}; \quad 0, q, 2q, \ldots, (p-1)q \tag{5.34}$$

consistute a difference set with parameters

$$v = qp, \quad k = (v-1)/N, \quad \lambda = (v-1-N)/N^2 \tag{5.35}$$

for g a common primitive root (easily provided by the Chinese Remainder Theorem) of the distinct odd primes p and q. Here $(p-1, q-1) = N$ and d is defined by $(p-1)(q-1) = dN$. Since $k = d + p = (v-1)/N$, it follows that $q = (N-1)p + 2$ is a necessary condition for such a difference set to exist; this condition is, in fact, sufficient when $N = 2$ (Theorem 5.27).

Whiteman approached this problem by developing a generalized cyclotomy for $v = pq$. In order to do this something akin to a primitive root must be established. Whiteman showed:

Lemma 5.28. Let g be a fixed common primitive root of both primes p and q; let $N = (p-1, q-1)$ and $dN = (p-1)(q-1)$. Then there exists an integer x such that the dN integers

$$g^s x^i \quad (s = 0, 1, \ldots, d-1; \quad i = 0, 1, \ldots, N-1) \tag{5.36}$$

constitute a reduced residue system modulo $v = pq$. [That is, all residues prime to v are of this form.]

Proof. Let x, y be a pair of integers satisfying the simultaneous congruences

$$\begin{aligned} x &\equiv g \pmod{p}, \quad y \equiv 1 \pmod{p} \\ x &\equiv 1 \pmod{q}, \quad y \equiv g \pmod{q}\,. \end{aligned} \tag{5.37}$$

The existence and uniqueness of such x, y are guaranteed by the Chinese

Remainder Theorem. Note that $xy \equiv g$ modulo v. Note further that the order of g modulo v is the least common multiple of $p - 1$, $q - 1$; i.e., it is d.

Now assume that $g^s x^i \equiv g^t x^j$, contrary to the lemma's assertion. Then (5.37) shows that $s \neq t$, while that fact that the order of g is d rules out the case $i = j$. So the assumption $g^s x^i \equiv g^t x^j$ can be written $x^\tau \equiv g^\sigma \equiv (xy)^\sigma$ with $\tau > 0$, $\sigma > 0$. By (5.37) this implies that $p - 1$ divides $\sigma - \tau$ and that $q - 1$ divides σ; so N divides τ. Since $0 < \tau < N$ this is a contradiction and the lemma has been established.

[Note that x is not unique, for y obviously serves equally as well.]

In Whiteman's generalized cyclotomy the <u>index class</u> i consists of the d numbers $(s = 0,1,\ldots,d - 1)$

$$a \equiv g^s x^i \qquad (\text{mod } v) \tag{5.38}$$

and the <u>generalized cyclotomic number</u> $(i,j)_N$ is the number of members of index class i which are followed by members of index class j. That is, $(i,j)_N$ is the number of solutions s, t of the congruence

$$g^s x^i + 1 \equiv g^t x^j \qquad (\text{mod } v) \tag{5.39}$$

where $0 \leq s, t \leq d - 1$.

Certain elementary properties of this generalized cyclotomy are needed for the difference set applications below. N is, of course, even and $p - 1 = Nf$, $q - 1 = Nf'$, $d = Nff'$ for some relatively prime integers f, f' (in particular f, f' are not both even).

$$x^N \equiv g^\mu \ (\text{mod } v) \quad \text{for some} \quad \mu \neq 1,\ 0 \leq \mu \leq d - 1 \tag{5.40}$$

$$-1 \equiv \begin{cases} g^{d/2} & (\text{mod } v) \quad \text{when} \quad ff' \text{ is odd} \\ g^\nu x^{N/2} & (\text{mod } v) \quad \text{when} \quad ff' \text{ is even} \end{cases} \tag{5.41}$$

where ν is some fixed integer, $0 \leq \nu \leq d - 1$.

$$(i,j)_N = (i',j')_N \quad \text{when} \quad i \equiv i', \ j \equiv j' \qquad (\text{mod } N) \tag{5.42}$$

$$(i,j)_N = (N - i, \ j - i)_N = \begin{cases} (j + N/2, \ i + N/2)_N & ff' \text{ even} \\ \\ (j,i)_N & ff' \text{ odd} \end{cases} \tag{5.43}$$

$$\sum_{j=0}^{N-1} (i,j)_N = \frac{(p - 2)(q - 2) - 1}{N} + \delta_i \tag{5.44}$$

where

$$\delta_i = \begin{cases} 1 & i \equiv N/2 \ (\text{mod } N) \quad ff' \text{ even} \\ 1 & i \equiv 0 \quad (\text{mod } N) \quad ff' \text{ odd} \\ 0 & \text{otherwise} \end{cases}$$

$$\sum_{j=0}^{N-1} (j,i) = \frac{(p - 2)(q - 2) - 1}{N} + \varepsilon_i \tag{5.45}$$

where

$$\varepsilon_i = \begin{cases} 1 & i \equiv 0 \ (N) \\ \\ 0 & \text{otherwise .} \end{cases}$$

Proof. If $x^N \neq g^\mu$ for any μ, then by (5.36) $x^N \equiv g^s x^i$ with $i > 0$. This implies $x^{N-i} \equiv_* g^s$ which contradicts Lemma 5.28. So $x^N \equiv g^\mu$ for some μ. If $\mu = 1$, then x is a primitive root of $v = pq$; this contradicts elementary number theory since p, q are distinct odd primes. So (5.40) has been established.

Suppose ff' is odd. Select x, y as in (5.37). Then $g^{d/2} \equiv x^{d/2}y^{d/2} \equiv -1$ modulo p as well as modulo q; hence $g^{d/2} \equiv -1$ mod v.

Let ff' be even and suppose $g^s \equiv -1$ modulo v. Then $x^s \equiv -1$ modulo p and $y^s \equiv -1$ modulo q; so s is divisible by $(p-1)/2$ as well as by $(q-1)/2$. Thus $d/2$ divides s and , as $s \neq 0$, $d/2$ is the only candidate. But $g^{d/2} \not\equiv -1$ modulo both p and q, since one of f, f' is even. So $-1 \equiv g^s x^i$ with $0 < i < N$. Thus $1 \equiv g^0 x^0 \equiv g^{2s} x^{2i}$ and it follows from Lemma 5.28 that $i = N/2$.

Equation (5.42) is an immediate consequence of the definition and the relation (5.40) above. The first part of (5.43) follows on multiplying (5.39) through by the inverse of its first term and using (5.40) to give

$$1 + g^{-s-\mu} x^{N-i} \equiv g^{t-s} x^{j-i} .$$

The second part of (5.43) follows from multiplying (5.39) by -1 as given in (5.41).

The summation if (5.44) counts, for fixed i, the numbers of the form $g^s x^i + 1$ which are in any index class at all. That is, all those $g^s x^i + 1$ which are prime to v. Of the d such numbers, let N_v be the number divisible by v, let N_p be the number divisible by p but not by q and let N_q be the number divisible by q but not by p. Then, clearly,

$$\sum_{j=0}^{N-1} (i,j)_N = d - N_v - N_p - N_q . \tag{5.46}$$

By (5.41) $N_v = \delta_i$. Since g is a primitive root modulo p it follows that all the non-zero residues of p are assumed $(q-1)/N$ times by g^s and hence by

$g^s x^i$ as s ranges from 0 to $d - 1$. Hence $N_p = (q - 1)/N - \delta_i$ and similarly $N_q = (p - 1)/N - \delta_i$. Equation (5.44) thus follows from 5.46.

Equation (5.45) follows when (5.43) is applied to (5.44). Thus all the relations above have been established.

Let the index class i be denoted by C_i.

Lemma 5.29. Let r be a fixed integer divisible by p or q but not by v. Then the number of solutions of the congruence

$$y - z \equiv r \pmod{v} \tag{5.47}$$

with y in class C_1 and z in class C_0 is independent of the value of r and is $(p - 1)(q - 1)/N^2$.

Proof. Let p divide r and let g, x generate the reduced residue system modulo v as in Lemma 5.28. Then $x \not\equiv 1 \pmod{v}$, and $g^u x \equiv 1 \pmod{p}$ for some fixed integer u such that $0 \leq u \leq p - 2$. In order for (5.47), i.e., in order for

$$g^t x - g^s \equiv r \pmod{v} \tag{5.48}$$

to be solvable, it is thus necessary that $t - s \equiv u \pmod{p - 1}$. Thus for each s $(0 \leq s \leq d - 1)$ there are precisely $(q - 1)/N$ values of t $[t = u + s + m(p-1)$, $0 \leq m < (q - 1)/N]$ for which the right side of (5.48) is divisible by p. Fix m and consider congruence 5.48 for any $q - 1$ consecutive values of s; the differences are $a, ga, \ldots, g^{q-2}a$. Since g is a primitive root of q and $a \not\equiv 0 \pmod{q}$, these are congruent modulo v to $p, 2p, \ldots, (q - 1)p$ in some order. Hence, for fixed m, as s ranges from s_0 to $s_0 + q - 2$ (5.48) represents any fixed r precisely once. For each value of m there are $d/(q - 1)$ such ranges of s. Thus a fixed r is represented exactly $(p - 1)(q - 1)/N^2$ times by (5.47). By symmetry the same result is true when q but not v divides r

and the lemma is proved.

Using these cyclotomic facts:

Theorem 5.30. (Whiteman, 1962) Let N denote the greatest common divisor of $p - 1$ $(=Nf)$ and $q - 1$ $(=Nf')$ where p and q are distinct odd primes, let $(p - 1)(q - 1) = dN$ and let g be a primitive root of both p and q. Then the numbers

$$1, g, g^2, \ldots, g^{d-1}; \quad 0, q, 2q, \ldots, (p - 1)q \tag{5.49}$$

form a difference set with parameters $v = pq$, $k = (v - 1)/N$, $\lambda = (v - 1 - N)/N^2$ if and only if ff' is even and the following two conditions are satisfied:

$$q = (N - 1)p + 2 \tag{5.50}$$

$$(i,0)_N = (N - 1)\,[(p - 1)/N]^2 \qquad (i = 0, 1, \ldots, N - 1)\,. \tag{5.51}$$

[Here $n = k - \lambda = (Nv - v + 1)/N^2 = (p - f)^2$ by 5.50 and so $(n,v) = 1$ for all difference sets of this type.]

The statement that the numbers (5.49) form a difference set is equivalent to the statement that for every fixed integer r, not divisible by v, there are precisely λ solutions to the congruence

$$y - z \equiv r \qquad (\mathrm{mod}\ v) \tag{5.52}$$

with y, z chosen from 5.49. The numbers $1, g, \ldots, g^{d-1}$ form the index class C_0 and the numbers $0, q, \ldots, (p - 1)q$ will be said to be members of the class Q. Before proving Theorem 5.30 a couple lemmas are required.

Lemma 5.31. Let r be a fixed integer not divisible by q. Then the number of solutions of congruence 5.52 with y in C_0 and z in Q is equal to $(p - 1)/N$.

Proof. As in the proof of Lemma 5.29 divide the integers $0, 1, \ldots, d - 1$ into $(p - 1)/N$ disjoint subsets, each of which contains $q - 1$ consecutive integers. Since g is a primitive root of q, there is exactly one number s in each of these subsets such that $g^s - r$ is divisible by q and the lemma follows.

Lemma 5.32. Let r be a fixed integer divisible by p but not by q. Then the number of solutions of congruence 5.52 with y and z both in class C_0 is $(p - 1)(q - 1 - N)/N^2$. When r is divisible by q but not by p the number of such solutions is $(q - 1)(p - 1 - N)/N^2$.

Proof. As in the proof of Lemma 5.29 a necessary condition for the solvability of $g^t - g^s \equiv r \pmod{v}$ is that $t \equiv s \pmod{p - 1}$. Thus this congruence becomes $g^{m(p-1)+s} - g^s \equiv r \pmod{v}$ with $1 \leq m < (q - 1)/N$. [The value $m = 0$ is excluded for then g divides r.] Fix m in this range and let s vary over $q - 1$ consecutive integers, then as before precisely one solution appears. So there are $(p - 1)(q - 1 - N)/N^2$ solutions overall, as was to be proved. Symmetry in p, q establishes the remainder of the lemma.

Proof of Theorem 5.30. Suppose that the numbers (5.49) constitute a difference set with the prescribed parameters. Then (5.50) is an immediate consequence of $k = d + p$.

Let r be relatively prime to v. Every solution of (5.52) with y and z in C_0 corresponds with a solution of $\bar{z}r + 1 \equiv \bar{z}y \pmod{v}$, here $\bar{z}z \equiv 1 \pmod{v}$. Thus if r belongs to C_i $(i = 0, \ldots, N - 1)$ the number of such solutions is $(i,0)_N$. Lemma 5.31 shows that if y belongs to C_0 and if z belongs to Q then there are $(p - 1)/N$ solutions to (5.52). By considering $-r$ instead, Lemma 5.31 also shows that there are $(p - 1)/N$ solutions with y in Q, z in C_0. Since $(r, v) = 1$, y and z cannot both be in Q and so the total number of solutions to (5.52) is

$$(i,0)_N + \frac{2(p - 1)}{N} = \lambda = \frac{v - 1 - N}{N^2}$$

from which (5.51) follows.

Suppose ff' were odd. Consider the cyclotomic number $(i,i)_N$; it counts the number of solutions of $r + 1 \equiv s \pmod{v}$ when r, s belong to C_i. Corresponding to any such solution is another solution $-s + 1 \equiv -r \pmod{v}$; since $-s$, $-r$ also belong to C_i (by 5.41). This second solution is distinct from the first unless $r \equiv -s \equiv -r - 1 \pmod{v}$, i.e., unless $v = 2r + 1$. Thus $(i,i)_N$ is even except when $(v - 1)/2$ belongs to C_i. Since, by (5.43), $(i,i)_N = (N - i,0)_N$, this implies that equation (5.51) cannot hold when ff' is odd. So ff' is even and the necessity part of the proof is complete.

Now suppose that conditions (5.50) and (5.51) hold; it is to be shown that congruence 5.52 has a uniform number of solutions independent of the value of r ($r \not\equiv 0$ modulo v) when y, z are chosen from among the numbers (5.49). There are three cases:

(i) Let p divide r. By Lemma 5.31 there are $(p - 1)/N$ solutions with y in Q and also $(p - 1)/N$ solutions with z in Q. Lemma 5.32 shows $(p - 1)(q - 1 - N)/N^2$ additional solutions when y, z are both in C_0. Using (5.50) the total number of solutions is $(v - 1 - N)/N^2$.

(ii) Let q divide r. When y, z both belong to C_0 Lemma 5.32 counts $(q - 1)(p - 1 - N)/N^2$ solutions. If y, z both belong to Q there are p solutions. So, by (5.50) the total number of solutions is $(v - 1 - N)/N^2$.

(iii) Let r be relatively prime to v. In the necessity part of the proof it was shown that the number of solutions in this case is $(i,0)_N + 2(p - 1)/N$ when r belongs to C_i. By (5.50), (5.51) this is again $(v - 1 - N)/N^2$.

So the numbers (5.49) do indeed form a difference set modulo $v = pq$ with $\lambda = (v - 1 - N)/N^2$ and $k = d + p = (v - 1)/N$. That is, Theorem 5.30 has been established.

As an example of this theorem consider the case $N = 2$. By (5.50) $q = p + 2$ and ff' is necessarily even. Equation (5.43) shows that $(0,0)_2 = (1,0)_2 = (1,1)_2$, so (5.51) is automatically satisfied. Thus the numbers (5.49) form a difference set whenever $q = p + 2$. That is, a proof has been given for Theorem 5.27 above. Of course, more direct proofs of this result exist (see the references given immediately below Theorem 5.27). With $p = 3$, $q = 5$, $g = 2$ then, the numbers 0, 1, 2, 4, 5, 8, 10 constitute a difference set with parameters $v, k, \lambda = 15, 7, 3$.

In applying Theorem 5.30 to the case $N = 4$ formulas for the cyclotomic numbers $(i,0)_4$ are needed. Whiteman (1962) derives these numbers. In particular he shows, for ff' even, that

$$\begin{aligned} 8(0,0)_4 &= -a + 2M + 3, & 8(0,1)_4 &= -a - 4b + 2M - 1 \\ 8(0,2)_4 &= 3a + 2M - 1, & 8(0,3)_4 &= -a + 4b + 2M - 1 \\ 8(1,0)_4 &= a + 2M + 1 & & \end{aligned} \tag{5.53}$$

the remaining $(i,j)_4$'s being equal to one of these as (5.43) shows. Here $M = [(p - 2)(q - 2) - 1)/4]$ and a, b are determined from a representation

$$v = a^2 + 4b^2, \qquad a \equiv 1 \qquad (\text{mod } 4) . \tag{5.54}$$

Whiteman observes that the common primitive roots of p, q can be divided into two classes G, G'. Here if g belongs to G then every primitive root in G is a power of g and similarly the primitive roots of G' are all powers of any particular one g'. There are two distinct representations of v as in (5.54). One such representation $a^2 + 4b^2$ is used with the primitive roots of G; the other, $a'^2 + 4b'^2$, with those of G'. Furthermore (Carlitz and Whiteman, 1964), the numbers $(0,0)_4$ and $(0,0)'_4$ computed for g and g' respectively are always unequal. In view of equation (5.51) this implies that at most one of g, g' gives rise to a difference set for $N = 4$.

Theorem 5.31. (Whiteman) Let p and q be distinct primes such that $(p - 1, q - 1) = 4$ and let $d = (p - 1)(q - 1)/4$. Let g, g' be distinct common primitive roots of p, q with $g' \not\equiv g^r \pmod{v}$ for any r. Then one (but not both) of the sets

$$1, g, g^2, \ldots, g^{d-1}; \quad 0, q, 2q, \ldots, (p - 1)q \tag{5.55}$$

$$1, g', g'^2, \ldots, g'^{d-1}; \quad 0, q, 2q, \ldots, (p - 1)q \tag{5.56}$$

is a difference set with parameters $v = pq$, $k = (v - 1)/4$, $\lambda = (v - 5)/16$ if and only if $q = 3p + 2$ and k is an odd square.

The generalized cyclotomic numbers $(i,j)_N$ have been computed for $N = 2,4$ (Whiteman, 1962) and for $N = 6,8$ (Bergquist, 1963). They may be found (ff' even only) together with a detailed discussion of the material of this section in Storer's booklet (1967 A). By way of summary, as far as difference sets are concerned, there is:

__Theorem 5.32.__ The numbers (5.49) form a difference set modulo $v = pq$ for $N = 2$ whenever $q = p + 2$ (Whiteman, 1962). For $N = 4$ (Whiteman, 1962) these numbers form a difference set for a suitable choice of the common primitive root g whenever $q = 3p + 2$ and k is an odd square. For $N = 6,8$ (Bergquist, 1963) these numbers never form a difference set.

For $N = 4$ the first example of such a difference set occurs with $v = pq = 901 = 17.53$ and the next example has $v = 6,575,588,101$.

Storer (1967 B) developed further cyclotomies along these lines and in particular succeeded in constructing the $v,k,\lambda = 133,33,8$ difference set of Hall (1956) by these methods.

Concerning the multiplier group of these difference sets and the multiplier conjecture there are the following facts:

__Theorem 5.33.__ The multiplier group of any difference set D given by (5.49) consists of the residues $1, g, \ldots, g^{d-1}$ and no others.

__Proof.__ These numbers are clearly multipliers. Suppose t belonging to index class j ($j \not\equiv 0$ modulo N) was also a multiplier. Then $td \equiv D + s$ (mod v) and so $s \not\equiv 0$ (mod v). Suppose q does not divide s. Since 0 belongs to D so does $-s$ and since $s \not\equiv 0 \pmod{q}$ it follows that $-s$ is a power of g. Thus (by 5.41) s belongs to index class $N/2$. Again since $s \not\equiv 0 \pmod{q}$, $tg^i - s \equiv 0$ for some i. So s and t as well as -1 belong to the same index class. Thus -1 is a multiplier if t is, contradiction.

Now consider the case where q divides s. Then, with t in C_j, $tD \equiv D + s$ implies that $C_0 + s \equiv C_j \pmod{v}$ with $s = q\ell$, $(s,p) = 1$. Since g is a primitive root modulo p, every non-zero residue modulo p occurs as g^i ranges over C_0. Thus $g^i + q\ell \equiv 0 \pmod{p}$ for some i, which contradicts $C_0 + s \equiv C_j$ and establishes the theorem.

Theorem 5.34. For $N = 2$ all divisors of n are multipliers for any difference set D given by (5.49).

Proof. When $N = 2$ the elements $1, g, \ldots, g^{d-1}$ are precisely those residues r, prime to v, for which $\left(\frac{r}{v}\right)$, the Jacobi symbol [see Nagell, 1951], is $+1$. Thus it is sufficient to show that every divisor d of n has $\left(\frac{d}{v}\right) = +1$. Using the reciprocity law for the Jacobi symbol the proof proceeds exactly like the proof of Theorem 5.20 above.

VI. MISCELLANY

In this chapter a few facts, some important, which was not convenient to mention earlier, are gathered.

A. Multiple Inequivalent Difference Sets

Singer (1938) showed that non-trivial difference sets with parameters

$$v = \frac{q^{N+1} - 1}{q - 1}, \quad k = \frac{q^{N} - 1}{q - 1}, \quad \lambda = \frac{q^{N-1} - 1}{q - 1} \tag{6.1}$$

exist whenever q is a prime power and $N \geq 2$; see section V.A. for a discussion of these difference sets. Singer noted that when $\lambda = 1$ there seemed to be only one equivalence class of difference sets with these parameters and he conjectured that this was in fact the case. This conjecture of Singer is still open. Hall (1956, p. 984) has verified Singer's conjecture for $q = 2,3,4,5,7,8,9,11,13,16,25,27,32$. Beyond this work of Hall apparently very little has been done on this problem.

Strangely enough the largest known class of multiple inequivalent difference sets also have parameters given by (6.1). Here, though, $N + 1$ is necessarily composite, thus $\lambda > 1$. In fact (see section V.A. for details) Gordon, Mills and Welch (1962) have shown that for any $\ell > 0$, there exist values of v, k, λ given by (6.1) for which there exist at least ℓ pairwise inequivalent difference sets.

For $k \leq 100$, $k < v/2$, the <u>known</u> multiple inequivalent difference sets are as follows: 2 for 31, 15, 7; 2 for 43, 21, 10; 2 for 63, 31, 15; 4 for 121, 40, 13 and 6 for 127, 63, 31. [There may well be others in this range. For while those parameters v, k, λ $(k \leq 100)$ which have associated cyclic difference sets have been determined (see section VI.B) not much beyond the work of Hall (1956, limited to $k \leq 50$) is known about multiple inequivalent difference sets in this range.]

Two block designs with the same parameters are considered equivalent if there exists two permutations (generally different), one acting on the objects and the other acting on the blocks, which take the one design into the other. On the other hand, equivalence of difference sets (see section I.B) is a more restrictive relation. Thus, it could happen that two cyclic difference sets were inequivalent while their associated block designs were not. There are no known examples of this behavior. In particular, it does not happen for the parameter sets $(k \leq 100)$ mentioned just above. In these cases, inequivalence of the associated block designs follows from the distribution of sizes of the intersections of triples of blocks. For example, the distribution of intersection size for all block triples containing a particular fixed block is 420 of size 3 and 15 of size 7 for one of the $v,k,\lambda = 31,15,7$ difference sets. Whereas the other 31, 15, 7 difference set has such block triples intersections 90 of size 2, 195 of size 3 and 150 of size 4. Since the distributions of block intersection sizes must be the same for equivalent block designs, it is clear that these designs are not equivalent.

The question of whether the multiple inequivalent difference sets of Gordon, Mills and Welch necessarily lead to inequivalent block designs is open.

[While no examples of inequivalent cyclic difference sets generating the same block design are known, there are examples of block designs which are generated by more than one difference set. Of course, at least one of the difference sets in these examples is non-cyclic. In fact, Bruck (1955, p. 475) has shown that the block design associated with a cyclic difference set with parameters $v,k,\lambda = n^2 + n + 1, n + 1, 1$ (these are only known to exist for prime power n) for $n \equiv 1$ modulo 3 can also be generated by a non-Abelian group difference set.]

B. Searches

In 1956 Marshall Hall (1956) published the results of his search for difference sets having $k \leq 50$. With but 12 exceptions, he decided, for each parameter set $v,k,\lambda (k \leq 50)$, whether or not a difference set existed. When the difference set was not a member of one of the families of Chapter V (and in some other cases) he listed the residues modulo v. For many of the smaller parameter sets he decided whether or not multiple inequivalent difference sets existed (see section VI.A. above).

Hall's twelve undecided cases were all resolved negatively. In fact many of the existence tests presented in section II.E. above were developed specifically for the purpose of deciding Hall's twelve cases. Utilizing these powerful tests Baumert (1969) extended Hall's search to $k \leq 100$. He found that there were exactly 74 parameter sets $v,k,\lambda (k \leq 100)$ for which non-trivial cyclic difference sets existed. These 74 parameter sets have associated with them 85 known difference sets; there may be more, since Baumert made no attempt to find multiple inequivalent difference sets beyond those already known. Table 6.1 of section VI.D. below contains all 85 known difference sets for which $k \leq 100$.

As reported in more detail in section IV.A., Hall (1947) checked that, for $n \leq 100$, all planar (i.e., $\lambda = 1$) difference sets had n a prime power. This conjecture, which is still outstanding, was checked further by Evans and Mann (1951) up to $n \leq 1600$. Dembowski (1968, p. 209) states that V. H. Keiser (unpublished) has checked it to $n \leq 3600$.

Hadamard difference sets (i.e., those with $v,k,\lambda = 4t - 1, 2t - 1, t - 1$) have been searched for, through $v < 1000$. The results of these searches (Golomb, Thoene, Baumert) are that, except for possible additional multiple inequivalent difference sets, the only unknown Hadamard difference sets possible, $v < 1000$, have $v = 399, 495, 627, 651, 783$ or 975. (See section IV.B. for a discussion of the known Hadamard difference sets.)

The difference sets associated with circulant Hadamard matrices (see section IV.C.) have been sought. These have parameters $v = 4N^2$, $n = N^2$. Turyn (1968) surveys the results in this area, most of which are due to him, and shows that, except for $v = 4$, none such exist with $v < 12,100$.

C. Some Examples

The general tenor of these examples seems to be that, at least from an algebraic number theoretic point of view, difference sets are no better behaved than they absolutely have to be.

Consider the quadratic residue difference set with parameters $v,k,\lambda = 103,51,25$. Here $n = 26$, the order of 2 modulo 103 is 51 and the order of 13 modulo 103 is 17, i.e., $2^{51} \equiv 1$, $13^{17} \equiv 1$ modulo 103, where the exponents are minimal. Furthermore, no power of 13 is congruent to 2 or 4 modulo 103. Thus Theorem 2.19 gives the prime ideal factorizations of 2 and 13 in the cyclotomic field $K(\zeta_{103})$ as:

$$(2) = Q\bar{Q}, \qquad (13) = P_1P_2P_3\bar{P}_1\bar{P}_2\bar{P}_3$$

where the bar denotes complex conjugation. Since $2^{48} \equiv 13$ it follows that the automorphism σ, defined by $\zeta_{103} \to \zeta_{103}^{13}$, fixes all these prime ideals; whereas the automorphism τ, defined by $\zeta_{103} \to \zeta_{103}^{2}$, fixes only the Q's. But τ^3, i.e., $\zeta_{103} \to \zeta_{103}^{8}$, fixes both P's and Q's, since $13^{16} \equiv 8 \pmod{103}$. On the other hand, Theorem 5.20 shows that 2 and 13 are both multipliers of this difference set. So, without loss of generality, the prime ideal decomposition of the ideal $(\theta(\zeta_{103}))$ is

$$(\theta(\zeta_{103})) = QP_1P_2P_3$$

where the action of the multiplier 2 permutes the prime ideal divisors of 13 according to $P_1 \to P_2 \to P_3 \to P_1$. That is, while the ideal $(\theta(\zeta_{103}))$ is fixed by the multiplier 2, its individual prime ideal divisors are not.

Consider the planar difference set whose parameters are $v,k,\lambda = 57,8,1$. It consists of the residues 1, 6, 7, 9, 19, 38, 42, 49 modulo 57. Using the notation of section III.D.,

$$\theta_{[57]}(x) = x + x^6 + x^7 + x^9 + x^{19} + x^{38} + x^{42} + x^{49}$$

$$\theta_{[19]}(x) = 2 + x + x^4 + x^6 + x^7 + x^9 + x^{11}$$

and the prime ideal divisors of $(\theta(\zeta_{19}))$ are (see Theorem 2.19):

$$A = (7,\ \zeta_{19}^3 + 6\zeta_{19}^2 + 3\zeta_{19} + 6)$$

$$B = (7,\ \zeta_{19}^3 + 5\zeta_{19}^2 + 6)$$

$$C = (7,\ \zeta_{19}^3 + 4\zeta_{19}^2 + 4\zeta_{19} + 6)$$

while $\overline{A}$, $\overline{B}$, $\overline{C}$ divide $(\theta(\zeta_{19}^{-1}))$. Each of these six ideals splits into two prime ideals over $K(\zeta_{57})$. Let a_1, a_2 and $\overline{b}_1$, $\overline{b}_2$ be the prime ideals of $K(\zeta_{57})$ which lie above A and $\overline{B}$ respectively. Then

$$(\theta(\zeta_{57})) = a_1\overline{a}_2 b_1\overline{b}_2 c_1 c_2$$

where

$$a_1 = (7,\ \zeta_{57}^3 + \zeta_{57}^2 + 2\zeta_{57} + 6)$$

$$\overline{a}_2 = (7,\ \zeta_{57}^3 + 6\zeta_{57}^2 + 5\zeta_{57} + 6)$$

$$b_1 = (7,\ \zeta_{57}^3 + 3\zeta_{57}^2 + 6\zeta_{57} + 6)$$

$$\overline{b}_2 = (7,\ \zeta_{57}^3 + 2\zeta_{57}^2 + 2\zeta_{57} + 6)$$

$$c_1 = (7,\ \zeta_{57}^3 + \zeta_{57} + 6)$$

$$c_2 = (7,\ \zeta_{57}^3 + 4\zeta_{57} + 6).$$

Thus, the tempting assumption that the fact that A divides the ideal $(\theta(\zeta_{19}))$ implies that one of a_1^2, a_2^2, a_1a_2 necessarily divides the corresponding ideal $(\theta(\zeta_{57}))$ is wrong. Besides certain theoretical implications, this complicates the application of the constructive existence test of section III.D.

Consider the planar difference set {1,2,4,8,16,32,37,55,64} whose parameters v, k, λ are 73, 9, 1; n = 8. Now one of the prime ideal divisors of (2) in $K(\zeta_{73})$ is

$$A = (2, \zeta_{73}^9 + \zeta_{73}^6 + \zeta_{73}^5 + \zeta_{73}^2 + 1)$$

and $A\bar{A}$ divides $(\theta(\zeta_{73}))$. Thus, the simplifying assumption, that one and only one of a complex conjugate pair of prime ideal divisors of n divides $(\theta(\zeta))$ is not correct in general.

D. A Table of Difference Sets

Table 6.1 below contains all 85 known difference sets for which $k \le 100$. As pointed out earlier, there well may be others; for the question of the existence of multiple inequivalent difference sets has not been solved for all of the 74 parameter sets v,k,λ ($k \le 100$) having associated difference sets. However, it has been shown that these 74 parameter sets are the only ones that need be considered.

Each difference set is identified by v, k, λ, n, p_1, p_2, p_3 (where the p_i are the prime divisors of n) and by a class indicator which indicates the family or sub-family to which it belongs. These are:

SN - Hyperplanes in projective N space (Theorem 5.1)
L - Quadratic Residues (Theorem 5.15)
B - Biquadratic Residues (Theorem 5.16)
BO - Biquadratic Residues and O (Theorem 5.18)
O - Octic Residues (Theorem 5.18)
H - Hall's Sets (Theorem 5.23)

TP - Twin Prime Sets (Theorem 5.27)

GMW - Gordon, Mills, Welch (Section V.A)

W - Whiteman Set (Theorem 5.31)

* - not one of the above .

Where multiple inequivalent sets are known, they are distinguished by A, B, C etc., which is added to the parameter v. Thus the two inequivalent $v,k,\lambda = 31,15,7$ difference sets are designated 31A, 15, 7 and 31B, 15, 7 respectively. The table is sorted in order of increasing values of k.

TABLE 6.1. DIFFERENCE SETS

V	K	λ	N	P1	P2							CLASS
7	3	1	2	2								S2=L
	*1	2	4									
13	4	1	3	3								S2=BO
	*0	1	3	9								
11	5	2	3	3								L
	*1	3	4	5	9							
21	5	1	4	2								S2
	*3	6	7	12	14							
31	6	1	5	5								S2
	*1	5	11	24	25	27						
15	7	3	4	2								S3=TP
	*0	1	2	4	5	8	10					
57	8	1	7	7								S2
	*1	6	7	9	19	38	42	49				
19	9	4	5	5								L
	*1	4	5	6	7	9	11	16	17			
37	9	2	7	7								B
	*1	7	9	10	12	16	26	33	34			
73	9	1	8	2								S2=B
	*1	2	4	8	16	32	37	55	64			
91	10	1	9	3								S2
	*0	1	3	9	27	49	56	61	77	81		
23	11	5	6	2	3							L
	*1	2	3	4	6	8	9	12	13	16	18	
133	12	1	11	11								S2
	*1	11	16	40	41	43	52	60	74	78	121	128

V	K	λ	N	P1	P2							CLASS								
40	13	4	9	3								S3								
	*1	2	3	5	6	9	14	15	18	20	25	27	35							
183	14	1	13	13								S2								
	*0	2	3	10	26	39	43	61	109	121	130	136	141	155						
31A	15	7	8	2								S4*H								
	*1	2	3	4	6	8	12	15	16	17	23	24	27	29	30					
31B	15	7	8	2								L								
	*1	2	4	5	7	8	9	10	14	16	18	19	20	25	28					
35	17	8	9	3								TP								
	*0	1	3	4	7	9	11	12	13	14	16	17	21	27	28	29	33			
273	17	1	16	2								S2								
	*1	2	4	8	16	32	64	91	117	128	137	182	195	205	234	239	256			
307	18	1	17	17								S2								
	*0	1	3	30	37	50	55	76	98	117	129	133	157	189	199	222	293	299		
381	20	1	19	19								S2								
	*0	1	19	28	96	118	151	153	176	202	240	254	290	296	300	307	337	361	366	369
43A	21	10	11	11								H								
	*1	2	3	4	5	8	11	12	16	19	20	21	22	27	32	33	35	37	39	41
	42																			
43B	21	10	11	11								L								
	*1	4	6	9	10	11	13	14	15	16	17	21	23	24	25	31	35	36	38	40
	41																			
85	21	5	16	2								S3								
	*0	1	2	4	7	8	14	16	17	23	27	28	32	34	43	46	51	54	56	64
	68																			
47	23	11	12	2	3							L								
	*1	2	3	4	6	7	8	9	12	14	16	17	18	21	24	25	27	28	32	34
	36	37	42																	
553	24	1	23	23								S2								
	*1	23	52	90	108	120	152	163	173	178	186	223	232	272	359	407	411	431	438	512
	513	515	529	548																
101	25	6	19	19								B								
	*1	5	16	19	24	25	31	36	37	52	54	56	58	68	71	78	79	80	81	84
	87	88	92	95	97															
651	26	1	25	5								S2								
	*1	5	25	42	71	107	125	201	210	217	354	355	357	387	399	412	434	462	468	473
	483	521	535	561	625	633														
109	28	7	21	3	7							B0								
	*0	1	3	5	7	9	15	16	21	22	25	26	27	35	38	45	48	49	63	66
	73	75	78	80	81	89	97	105												

∇	K	λ	N	P1	P2							CLASS								
757	28	1	27	3								S2								
	*0	1	3	9	27	43	81	129	173	220	243	310	387	404	409	445	455	466	470	505
	519	578	608	641	653	660	673	729												
59	29	14	15	3	5							L								
	*1	3	4	5	7	9	12	15	16	17	19	20	21	22	25	26	27	28	29	35
	36	41	45	46	48	49	51	53	57											
871	30	1	29	29								S2								
	*1	24	29	69	151	167	216	234	259	263	295	321	329	414	488	543	582	599	645	659
	683	689	696	716	731	819	820	822	831	841										
63A	31	15	16	2								S5								
	*0	1	2	3	4	6	7	8	9	12	13	14	16	18	19	24	26	27	28	32
	33	35	36	38	41	45	48	49	52	54	56									
63B	31	15	16	2								GMW								
	*0	1	2	3	4	5	6	8	9	10	12	16	17	18	20	23	24	27	29	32
	33	34	36	40	43	45	46	48	53	54	58									
156	31	6	25	5								S3								
	*0	1	5	11	13	25	28	39	46	55	58	65	68	74	76	86	87	91	111	117
	118	119	122	123	125	127	134	140	142	143	147									
993	32	1	31	31								S2								
	*0	1	23	31	60	66	77	84	87	195	253	257	291	331	401	416	468	473	515	590
	606	618	711	713	752	761	841	867	892	912	961	980								
67	33	16	17	17								L								
	*1	4	6	9	10	14	15	16	17	19	21	22	23	24	25	26	29	33	35	36
	37	39	40	47	49	54	55	56	59	60	62	64	65							
133	33	8	25	5								*								
	*1	4	5	14	16	19	20	21	25	38	54	56	57	64	66	70	76	80	83	84
	91	93	95	98	100	101	105	106	114	123	125	126	131							
1057	33	1	32	2								S2								
	*1	2	4	8	16	32	55	64	110	128	139	220	256	278	299	339	349	440	453	512
	529	556	598	678	698	703	755	793	880	906	925	991	1024							
71	35	17	18	2	3							L								
	*1	2	3	4	5	6	8	9	10	12	15	16	18	19	20	24	25	27	29	30
	32	36	37	38	40	43	45	48	49	50	54	57	58	60	64					
1407	38	1	37	37								S2								
	*0	1	37	63	205	274	289	302	314	316	321	362	414	420	436	465	469	486	550	621
	644	652	655	711	731	844	854	924	981	1098	1122	1152	1187	1230	1248	1316	1325	1369		
79	39	19	20	2	5							L								
	*1	2	4	5	8	9	10	11	13	16	18	19	20	21	22	23	25	26	31	32
	36	38	40	42	44	45	46	49	50	51	52	55	62	64	65	67	72	73	76	

V	K	λ	N	P1	P2							CLASS								
121A	40	13	27	3								S4								
	*1	3	4	7	9	11	12	13	21	25	27	33	34	36	39	44	55	63	64	67
	68	70	71	75	80	81	82	83	85	89	92	99	102	103	104	108	109	115	117	119
121B	40	13	27	3								*								
	*1	3	4	5	9	12	13	14	15	16	17	22	23	27	32	34	36	39	42	45
	46	48	51	64	66	69	71	77	81	82	85	86	88	92	96	102	108	109	110	117
121C	40	13	27	3								*								
	*1	3	4	7	8	9	12	21	24	25	26	27	34	36	40	43	49	63	64	68
	70	71	72	75	78	81	82	83	89	92	94	95	97	102	104	108	112	113	118	120
121D	40	13	27	3								*								
	*1	3	4	5	7	9	12	14	15	17	21	27	32	36	38	42	45	46	51	53
	58	63	67	68	76	79	80	81	82	83	96	100	103	106	107	108	114	115	116	119
83	41	20	21	3	7							L								
	*1	3	4	7	9	10	11	12	16	17	21	23	25	26	27	28	29	30	31	33
	36	37	38	40	41	44	48	49	51	59	61	63	64	65	68	69	70	75	77	78
	81																			
1723	42	1	41	41								S2								
	*1	31	41	99	152	187	232	271	313	345	361	421	508	594	613	618	638	672	679	710
	761	772	773	775	879	897	988	996	1011	1017	1063	1067	1194	1207	1216	1243	1271	1542	1579	1612
	1681	1707																		
1893	44	1	43	43								S2								
	*0	1	5	43	147	164	215	284	308	356	375	439	457	459	537	627	642	655	715	721
	744	755	777	780	807	854	981	1104	1230	1262	1338	1359	1373	1468	1507	1592	1647	1663	1673	1704
	1779	1840	1849	1886																
2257	48	1	47	47								S2								
	*1	47	102	136	149	163	232	243	263	280	353	572	622	792	848	890	918	994	999	1037
	1076	1112	1136	1204	1289	1324	1342	1375	1429	1481	1487	1578	1702	1710	1742	1813	1875	1876	1878	1885
	1897	1901	1942	2057	2135	2150	2179	2209												
197	49	12	37	37								B								
	*1	16	23	24	28	29	34	36	37	40	42	49	51	53	54	59	60	61	63	70
	76	81	85	88	90	100	101	104	105	114	132	133	135	142	150	154	156	158	164	171
	172	175	178	182	187	188	190	191	193											
2451	50	1	49	7								S2								
	*0	1	7	49	170	223	267	343	348	382	491	508	688	750	828	886	894	977	986	989
	1105	1123	1177	1190	1300	1304	1356	1561	1634	1716	1745	1747	1775	1849	1869	1892	1937	2000	2021	2101
	2139	2171	2208	2269	2346	2365	2401	2411	2425	2436										
103	51	25	26	2	13							L								
	*1	2	4	7	8	9	13	14	15	16	17	18	19	23	25	26	28	29	30	32
	33	34	36	38	41	46	49	50	52	55	56	58	59	60	61	63	64	66	68	72
	76	79	81	82	83	91	92	93	97	98	100									

V	K	λ	N	P1	P2							CLASS								
107	53	26	27	3								L								
	*1	3	4	9	10	11	12	13	14	16	19	23	25	27	29	30	33	34	35	36
	37	39	40	41	42	44	47	48	49	52	53	56	57	61	62	64	69	75	76	79
	81	83	85	86	87	89	90	92	99	100	101	102	105							
2863	54	1	53	53								S2								
	*1	21	53	72	189	225	292	345	386	417	443	451	473	561	566	575	644	904	929	953
	966	999	1103	1107	1113	1161	1199	1234	1246	1347	1368	1410	1411	1413	1428	1487	1510	1700	1729	1838
	1845	2060	2076	2104	2165	2233	2416	2443	2527	2639	2679	2718	2729	2809						
400	57	8	49	7								S3								
	*1	7	25	34	49	52	54	62	66	68	75	76	103	106	109	122	124	125	129	132
	141	148	155	169	170	175	187	200	211	225	236	238	246	247	275	277	281	285	310	321
	325	330	339	342	343	358	363	364	365	367	373	375	378	383	390	394	395			
3541	60	1	59	59								S2								
	*1	59	63	79	102	135	176	266	348	366	373	374	376	462	595	608	629	761	820	883
	938	962	1009	1120	1211	1379	1421	1530	1557	1607	1701	1713	1745	1850	1919	2187	2227	2244	2312	2342
	2347	2396	2407	2471	2477	2523	2728	2747	2827	2875	2920	3198	3236	3251	3265	3302	3338	3450	3459	3481
3783	62	1	61	61								S2								
	*0	1	61	220	239	246	268	314	400	570	572	613	641	650	720	723	756	845	950	1042
	1107	1205	1216	1271	1313	1347	1404	1488	1492	1628	1681	1702	1820	1836	1871	2071	2161	2206	2289	2299
	2307	2319	2366	2418	2490	2522	2724	3034	3199	3216	3230	3243	3346	3441	3490	3495	3607	3657	3663	3721
	3744	3759																		
127A	63	31	32	2								L								
	*1	2	4	8	9	11	13	15	16	17	18	19	21	22	25	26	30	31	32	34
	35	36	37	38	41	42	44	47	49	50	52	60	61	62	64	68	69	70	71	72
	73	74	76	79	81	82	84	87	88	94	98	99	100	103	104	107	113	115	117	120
	121	122	124																	
127B	63	31	32	2								F								
	*1	2	3	4	5	6	7	8	10	12	14	16	19	20	23	24	25	27	28	32
	33	38	40	46	47	48	50	51	54	56	57	61	63	64	65	66	67	73	75	76
	77	80	87	89	92	94	95	96	97	100	101	102	107	108	111	112	114	117	119	122
	123	125	126																	
127C	63	31	32	2								S6								
	*1	2	3	4	6	7	8	9	12	14	15	16	17	18	24	27	28	29	30	31
	32	34	36	39	47	48	51	54	56	58	60	61	62	64	65	67	68	71	72	77
	78	79	83	87	89	94	96	97	99	102	103	105	107	108	112	113	115	116	117	120
	121	122	124																	
127D	63	31	32	2								*								
	*1	2	3	4	6	7	8	9	12	13	14	16	17	18	19	24	25	26	27	28
	31	32	34	35	36	38	47	48	50	51	52	54	56	61	62	64	65	67	68	70
	72	73	76	77	79	81	87	89	94	96	97	100	102	103	104	107	108	112	115	117
	121	122	124																	

V	K	λ	N	P1	P2						CLASS									
127F	63	31	32	2							*									
	*1	2	3	4	5	6	8	9	10	12	15	16	17	18	19	20	24	25	27	29
	30	32	33	34	36	38	39	40	48	50	51	54	55	58	59	60	64	65	66	68
	71	72	73	76	77	78	80	83	89	91	93	96	99	100	102	105	108	109	110	113
	116	118	120																	
127F	63	31	32	2							*									
	*1	2	3	4	5	6	8	10	11	12	16	19	20	21	22	24	25	27	29	32
	33	37	38	39	40	41	42	44	48	49	50	51	54	58	63	64	65	66	69	73
	74	76	77	78	80	82	83	84	88	89	95	96	98	100	102	105	108	111	116	119
	123	125	126																	
131	65	32	33	3	11						L									
	*1	3	4	5	7	9	11	12	13	15	16	20	21	25	27	28	33	34	35	36
	38	39	41	43	44	45	46	48	49	52	53	55	58	59	60	61	62	63	64	65
	74	75	77	80	81	84	89	91	94	99	100	101	102	105	107	108	109	112	113	114
	117	121	123	125	129															
4161	65	1	64	2							S2									
	*1	2	4	8	16	32	64	128	256	285	307	357	399	425	512	570	614	714	751	798
	850	1024	1117	1140	1228	1387	1428	1502	1551	1596	1605	1700	1761	1847	2043	2048	2081	2223	2234	2259
	2280	2293	2456	2639	2774	2856	2883	2961	3004	3102	3121	3192	3210	3227	3400	3522	3561	3641	3694	3861
	3901	4011	4031	4086	4096															
4557	68	1	67	67							S2									
	*0	1	25	67	192	213	256	305	328	350	363	481	491	544	600	615	665	820	858	897
	998	1150	1318	1359	1516	1519	1536	1675	1723	1760	2045	2102	2107	2153	2207	2548	2629	2658	2762	2774
	2802	2857	2888	2904	2977	2984	3036	3068	3174	3285	3359	3481	3508	3542	3578	3744	3748	3750	3826	3977
	3995	4021	4124	4138	4459	4470	4489	4549												
139	69	34	35	5	7						L									
	*1	4	5	6	7	9	11	13	16	20	24	25	28	29	30	31	34	35	36	37
	38	41	42	44	45	46	47	49	51	52	54	55	57	63	64	65	66	67	69	71
	77	78	79	80	81	83	86	89	91	96	99	100	106	107	112	113	116	117	118	120
	121	122	124	125	127	129	131	136	137											
143	71	35	36	2	3						TP									
	*0	1	2	3	4	6	7	8	9	12	13	14	16	18	19	21	23	24	25	26
	27	28	32	36	38	39	41	42	46	48	49	50	52	53	54	56	57	63	64	65
	69	72	73	75	76	78	81	82	83	84	85	91	92	96	98	100	103	104	106	108
	109	112	113	114	117	123	126	128	130	133	138									
5113	72	1	71	71							S2									
	*1	71	222	260	419	423	499	510	819	829	877	897	911	1179	1361	1502	1668	1707	1732	1885
	1901	1906	2003	2004	2006	2033	2143	2331	2346	2388	2435	2580	2616	2704	2803	2856	2869	2931	2950	3065
	3121	3325	3369	3421	3581	3598	3635	3714	3845	3876	3989	4001	4061	4156	4162	4184	4207	4225	4233	4268
	4292	4342	4375	4382	4468	4597	4719	4751	4921	4930	4976	5041								

V	K	λ	N	P1	P2	P3						CLASS								
585	73	9	64	2								S3								
	*0	1	2	4	5	8	10	16	17	20	32	34	39	40	43	55	64	65	68	78
	80	86	103	110	123	128	130	136	156	157	160	172	177	195	206	213	220	239	246	256
	257	260	267	272	293	295	301	312	314	320	325	344	354	371	381	390	399	412	421	426
	439	440	443	455	478	483	492	503	512	514	520	534	544							
5403	74	1	73	73								S2								
	*0	1	23	73	275	352	470	546	644	686	732	734	864	885	900	967	971	979	985	1104
	1148	1228	1262	1291	1451	1470	1496	1535	1666	1679	1720	1801	1892	2037	2392	2427	2703	2728	2752	2759
	2811	2820	3041	3164	3196	3266	3306	3442	3596	3606	3639	3701	3788	3866	3894	3995	4046	4084	4104	4275
	4636	4653	4683	4749	4752	4809	4950	4955	5117	5172	5265	5276	5292	5329						
151	75	37	38	2	19							L								
	*1	2	4	5	8	9	10	11	16	17	18	19	20	21	22	25	29	31	32	34
	36	37	38	39	40	42	43	44	45	47	49	50	55	58	59	62	64	68	69	72
	74	76	78	80	81	84	85	86	88	90	91	94	95	97	98	99	100	103	105	110
	116	118	121	123	124	125	127	128	136	137	138	139	144	145	148					
6321	80	1	79	79								S2								
	*0	1	14	17	43	79	255	490	578	717	784	822	993	1037	1106	1143	1182	1324	1343	1415
	1479	1537	1728	1779	1803	1836	1906	1989	2375	2384	2540	2595	2712	2733	2783	2785	2881	3007	3014	3063
	3246	3375	3397	3460	3594	3676	3771	4214	4229	4275	4316	4328	4709	4756	4824	4884	4916	4943	4961	5027
	5047	5101	5191	5201	5226	5231	5393	5399	5427	5534	5545	5655	5802	5850	5951	5959	5982	6071	6075	6241
163	81	40	41	41								L								
	*1	4	6	9	10	14	15	16	21	22	24	25	26	33	34	35	36	38	39	40
	41	43	46	47	49	51	53	54	55	56	57	58	60	61	62	64	65	69	71	74
	77	81	83	84	85	87	88	90	91	93	95	96	97	100	104	111	113	115	118	119
	121	126	131	132	133	134	135	136	140	143	144	145	146	150	151	152	155	156	158	160
	161																			
6643	82	1	81	3								S2								
	*0	1	3	9	27	32	81	96	151	199	243	288	453	457	509	584	597	614	729	803
	864	904	932	1003	1133	1359	1371	1493	1527	1745	1752	1791	1842	1856	2044	2187	2225	2384	2409	2419
	2482	2525	2592	2712	2796	2833	2956	3009	3056	3233	3292	3399	3418	3478	3554	3611	3791	3802	4019	4077
	4113	4190	4429	4479	4495	4581	4730	4763	5110	5235	5256	5373	5414	5526	5568	5588	5696	5905	5927	6132
	6397	6561																		
167	83	41	42	2	3	7						L								
	*1	2	3	4	6	7	8	9	11	12	14	16	18	19	21	22	24	25	27	28
	29	31	32	33	36	38	42	44	47	48	49	50	54	56	57	58	61	62	63	64
	65	66	72	75	76	77	81	84	85	87	88	89	93	94	96	97	98	99	100	107
	108	112	114	115	116	121	122	124	126	127	128	130	132	133	137	141	144	147	150	152
	154	157	162																	

V	K	λ	N	P1	P2							CLASS								
6973	84	1	83	83								S2								
	*1	83	219	242	274	275	277	293	472	591	598	691	699	823	900	1043	1056	1103	1129	1201
	1544	1569	1635	1823	1906	1945	2044	2061	2072	2120	2190	2233	2300	2508	2523	2629	2638	2786	2791	2893
	3037	3058	3218	3280	3400	3429	3504	3599	3677	3687	3691	3711	3741	3927	3972	4041	4077	4174	4231	4311
	4496	4624	4713	4765	4792	4830	4836	4876	4917	4939	4970	5007	5183	5352	5491	5503	5552	5687	5851	5947
	6140	6182	6514	6889																
341	85	21	64	2								S4								
	*1	2	4	5	8	10	13	16	20	23	25	26	27	32	39	40	46	50	52	54
	55	59	64	71	75	78	80	91	92	95	99	100	103	104	108	109	110	111	113	118
	121	128	131	142	143	150	156	160	165	171	173	177	182	183	184	190	198	200	206	208
	216	218	220	222	225	226	227	231	236	242	253	256	257	259	262	283	284	286	297	299
	300	312	319	320	330															
179	89	44	45	3	5							L								
	*1	3	4	5	9	12	13	14	15	16	17	19	20	22	25	27	29	31	36	39
	42	43	45	46	47	48	49	51	52	56	57	59	60	61	64	65	66	67	68	70
	74	75	76	77	80	81	82	83	85	87	88	89	93	95	100	101	106	107	108	110
	116	117	121	124	125	126	129	135	138	139	141	142	144	145	146	147	149	151	153	155
	156	158	161	168	169	171	172	173	177											
8011	90	1	89	89								S2								
	*1	50	82	89	341	342	344	403	411	631	647	855	927	931	1024	1043	1084	1264	1300	1354
	1382	1407	1506	1546	1572	1789	1914	2115	2232	2262	2373	2393	2598	2718	2727	2749	2833	2911	3015	3065
	3148	3160	3165	3511	3546	3721	3785	3796	3823	3917	3959	3972	3982	3996	4030	4140	4185	4331	4387	4450
	4535	4564	4691	4706	4828	5058	5076	5109	5646	5720	5806	5812	5858	5915	6085	6186	6316	6384	6405	6510
	6583	6914	7012	7221	7298	7406	7798	7878	7921	7965										
820	91	10	81	3								S3								
	*1	3	9	17	22	26	27	31	35	41	51	53	64	66	78	81	88	93	94	95
	97	105	123	125	142	153	159	178	192	193	198	202	205	234	236	239	243	256	264	279
	282	285	287	291	305	315	352	353	369	375	391	410	426	451	458	459	466	477	478	484
	499	511	533	534	547	554	557	568	576	578	579	594	606	611	614	615	632	664	677	697
	702	706	708	713	717	729	736	768	779	782	792									
191	95	47	48	2	3							L								
	*1	2	3	4	5	6	8	9	10	12	13	15	16	17	18	20	23	24	25	26
	27	30	32	34	36	39	40	43	45	46	48	49	50	51	52	54	59	60	64	65
	67	68	69	72	75	77	78	79	80	81	85	86	90	92	96	97	98	100	102	103
	104	107	108	109	115	117	118	120	121	125	128	129	130	133	134	135	136	138	144	147
	149	150	153	154	156	158	160	162	163	169	170	172	177	180	184					

V	K	λ	N	P1	P2							CLASS								
9507	98	1	97	97								S2								
	*1	13	68	97	137	360	568	611	657	670	696	717	833	889	963	1070	1071	1073	1107	1122
	1261	1378	1402	1503	1984	1989	2054	2163	2225	2301	2308	2670	2748	2793	2802	2825	2843	2896	3000	3008
	3169	3186	3211	3527	3782	3929	4128	4257	4536	4594	4725	4745	4818	5209	5215	5253	5367	5371	5588	5598
	5670	5790	5847	6034	6113	6124	6246	6338	6399	6426	6566	6596	6671	6687	6921	7221	7243	7561	7609	7829
	7848	7862	7948	8091	8233	8296	8360	8560	8720	8817	8930	9011	9098	9126	9224	9374	9409	9440		
199	99	49	50	2	5							L								
	*1	2	4	5	7	8	9	10	13	14	16	18	20	23	25	26	28	29	31	32
	33	35	36	40	43	45	46	47	49	50	51	52	53	56	57	58	61	62	63	64
	65	66	70	72	79	80	81	86	89	90	91	92	94	98	100	102	103	104	106	111
	112	114	115	116	117	121	122	123	124	125	126	128	130	131	132	139	140	144	145	151
	155	157	158	160	161	162	165	169	172	175	177	178	180	182	184	187	188	193	196	
323	161	80	81	3								TP								
	*0	1	3	4	9	10	12	14	16	19	22	25	26	27	29	30	31	35	36	37
	38	40	41	42	43	46	47	48	49	55	56	57	64	65	66	71	75	76	77	78
	79	81	83	87	88	90	91	93	95	97	100	101	104	105	106	107	108	109	111	113
	114	115	116	118	120	121	122	123	124	126	129	133	134	137	138	140	141	143	144	146
	147	148	149	152	157	160	161	164	165	167	168	169	171	172	173	178	181	184	188	190
	191	192	193	195	196	198	206	209	211	213	220	224	225	227	228	229	231	234	237	239
	241	243	247	249	250	251	253	254	256	260	261	262	263	264	265	266	269	270	271	273
	278	279	284	285	290	291	295	299	300	302	303	304	305	308	310	312	315	316	317	318
	321																			
901	225	56	169	13								W								
	*0	1	5	9	12	13	14	16	22	25	41	43	45	47	53	59	60	65	69	70
	71	79	80	81	89	92	93	106	108	109	110	114	117	124	125	126	133	139	144	147
	152	156	159	167	168	169	173	174	182	183	192	194	196	198	202	203	205	208	209	212
	214	215	219	222	223	224	225	226	229	231	232	233	235	244	254	256	259	264	265	274
	277	286	292	293	295	296	300	307	308	313	318	319	325	326	345	350	352	355	363	369
	371	379	382	387	394	395	397	400	401	402	405	407	419	422	423	424	433	445	447	460
	461	465	467	469	477	484	492	498	502	503	516	523	526	529	530	531	533	536	540	543
	545	550	559	564	570	571	574	577	579	581	583	585	587	596	599	602	611	617	618	620
	621	622	625	630	634	636	639	641	656	658	661	664	665	688	689	691	694	695	706	708
	711	713	720	721	724	729	735	737	742	746	752	760	766	767	772	778	780	786	795	801
	813	824	826	827	828	835	837	840	843	845	848	849	852	853	859	862	863	865	870	874
	878	881	886	897	898															

REFERENCES

Generally speaking a survey such as this is expected to contain a fairly exhaustive bibliography. However, since the book of Dembowski (1968) contains an excellent 49 page bibliography covering this subject, the listing here may be, and is, restricted to items explicitly mentioned in this report.

Alanen, J. D., and D. E. Knuth

1964 "Tables of Finite Fields", Sankhyā 26, 305-328.

Albert, A. A.

1953 "Rational Normal Matrices Satisfying the Incidence Equation", Proc. Amer. Math. Soc. 4, 554-559.

Albert, A. A., and R. Sandler

1968 An Introduction to Finite Projective Planes, Holt, Rinehart and Winston, New York

Barker, R. H.

1953 "Group Synchronizing of Binary Digital Systems", in Communication Theory, W. Jackson, Editor, London, 273-287.

Baumert, L. D.

1969 "Difference Sets", SIAM J. Appl. Math. 17, 826-833.

Baumert, L. D., and H. Fredricksen

1967 "The Cyclotomic Numbers of Order Eighteen with Applications to Difference Sets", Math. Comp. 21, 204-219.

Belevitch, V.

1968 "Conference Networks and Hadamard Matrices", Ann. Soc. Sci. Bruxelles Ser. I 82, 13-32.

Bergquist, J. W.

1963 "Difference Sets and Congruences Modulo a Product of Primes", Dissertation, University of Southern California.

Brauer, A.

1953 "On a New Class of Hadamard Determinants", Math. Z. 58, 219-225.

Brualdi, R.A.

1965 "A Note on Multipliers of Difference Sets", J. Res. Nat. Bur. Standards Sect. B 69B, 87-89.

Bruck, R. H.

1955 "Difference Sets in a Finite Group", Trans. Amer. Math. Soc. 78, 464-481.

Bruck, R. H., and H. J. Ryser

1949 "The Nonexistence of Certain Finite Projective Planes", Canad. J. Math. 1, 88-93.

Carlitz, L., and A. L. Whiteman

1964 "The Number of Solutions of Some Congruences Modulo a Product of Primes", Trans. Amer. Math. Soc. 112, 536-552.

Chowla, S.

1944 "A Property of Biquadratic Residues", Proc. Nat. Acad. Sci. India, Sect. A 14, 45-46.

1945 "Contributions to the Theory of the Construction of Balanced Incomplete Block Designs", Math. Student 12, 82-85.

Chowla, S., and H. J. Ryser

1950 "Combinatorial Problems", Canad. J. Math. 2, 93-99.

Dembowski, P.

1968 Finite Geometries, Springer-Verlag, New York.

Dickson, L. E.

1935A "Cyclotomy, Higher Congruences and Waring's Problem", Amer. J. Math. 57, 391-424 and 463-474.

1935B "Cyclotomy and Trinomial Congruences", Trans. Amer. Math. Soc. 37, 363-380.

1935C "Cyclotomy When E is Composite", Trans. Amer. Math. Soc. 38, 187-200.

Evans, T. A., and H. B. Mann

1951 "On Simple Difference Sets", Sankhyā 11, 357-364.

Gillies, D. B.

1964 "Three New Mersenne Primes and a Statistical Theory", Math. Comp. 18, 93-95.

Goethals, J. M., and J. J. Seidel

1967 "Orthogonal Matrices with Zero Diagonal", Canad. J. Math. 19, 1001-1010.

Goldhaber, J. K.

1960 "Integral P-Adic Normal Matrices Satisfying the Incidence Equation", Canad. J. Math. 12, 126-133.

Goldstein, R. M.

1964 "Venus Characteristics by Earth-Based Radar", Astronom. J. 69, 12-18.

Golomb, S. W. et al.

1964 Digital Communications with Space Applications, Prentice-Hall, Englewood Cliffs, New Jersey.

Gordon, B., W. H. Mills and L. R. Welch

1962 "Some New Difference Sets", Canad. J. Math. 14, 614-625.

Gruner, W.

1939 "Einlagerung des Regularen N-Simplex in den N-Dimensionalen Wurfel", Comment. Math. Helv. 12, 149-152.

Halberstam, H., and R. R. Laxton

1964 "Perfect Difference Sets", Proc. Glasgow Math. Assoc. 6, 177-184.

Hall, M. Jr.

1947 "Cyclic Projective Planes", Duke Math. J. 14, 1079-1090.

1956 "A Survey of Difference Sets", Proc. Amer. Math. Soc. 7, 975-986.

1965 "Characters and Cyclotomy", Proc. Sym. in Pure Math., Amer. Math. Soc. 8, 31-43.

1967 Combinatorial Theory, Blaisdell, Waltham, Massachusetts.

Hall, M. Jr., and H. J. Ryser

1951 "Cyclic Incidence Matrices", Canad. J. Math. 3, 495-502.

1954 "Normal Completions of Incidence Matrices", Amer. J. Math. 76, 581-589.

Hayashi, H. S.

1965 "Computer Investigation of Difference Sets", Math. Comp. 19, 73-78.

Hughes, D. R.

1957 "Regular Collineation Groups", Proc. Amer. Math. Soc. 8, 165-168.

Johnsen, E. C.

1964 "The Inverse Multiplier for Abelian Group Difference Sets", Canad. J. Math. 16, 787-796.

1965 "Matrix Rational Completions Satisfying Generalized Incidence Equations", Canad. J. Math. 17, 1-12.

1966A "Integral Solutions to the Incidence Equation for Finite Projective Plane Cases of Orders $N \equiv 2 \pmod 4$", Pacific J. Math. 17, 97-120.

1966B "Skew-Hadamard Abelian Group Difference Sets", J. Algebra, 388-402.

Jones, B. W.

1950 The Arithmetic Theory of Quadratic Forms, Carus Math. Mono. 10, Wiley, New York.

Kesava Menon, P.

1962 "Certain Hadamard Designs", Proc. Amer. Math. Soc. 13, 524-531.

Lebesgue, A. U.

1850 "Sur L'Impossibilite, en Nombres Entiers, de L'Equation $x^m = y^2 + 1$", Nouvelles Ann. Math. 9, 178-181.

Lehmer, E.

1953 "On Residue Difference Sets", Canad. J. Math. 5, 425-432.

1955A "Period Equations Applied to Difference Sets", Proc. Amer. Math. Soc. 6, 433-442.

1955B "On the Number of Solutions of $u^k + D \equiv w^2 \pmod p$", Pacific J. Math. 5, 103-118.

Magnus, W.

1937 "Uber die Anzahl der in Einem Geschlecht Enthaltenen Klassen von Positiv Definiten Quadratischen Formen", Math. Ann. 114, 465-475.

Mann, H. B.

1952 "Some Theorems on Difference Sets", Canad. J. Math. 4, 222-226.

1955 Introduction to Algebraic Number Theory, Ohio State Univ. Press, Columbus, Ohio.

Mann, H. B.

1964 "Balanced Incomplete Block Designs and Abelian Difference Sets", Illinois J. Math. 8, 252-261.

1965 Addition Theorems, The Addition Theorems of Group Theory and Number Theory, Interscience, New York.

1967 "Recent Advances in Difference Sets", Amer. Math. Monthly 74, 229-235.

Mann, H. B., and S. K. Zaremba

1969 "On Multipliers of Difference Sets", Illinois J. Math. 13, 378-382.

McFarland, R., and H. B. Mann

1965 "On Multipliers of Difference Sets", Canad. J. Math. 17, 541-542.

Muskat, J. B.

1966 "The Cyclotomic Numbers of Order Fourteen", Acta Arith. 11, 263-279.

Nagell, T.

1951 Introduction to Number Theory, Wiley, New York.

Neumann, H.

1955 "On Some Finite Non-Desarguesian Planes", Arch. Math. 6, 36-40.

Newman, M.

1963 "Multipliers of Difference Sets", Canad. J. Math. 15, 121-124.

O'Meara, O. T.

1963 Introduction to Quadratic Forms, Springer-Verlag, New York.

Ostrom, T. G.

1953 "Concerning Difference Sets", Canad. J. Math. 5, 421-424.

Paley, R. E. A. C.

1933 "On Orthogonal Matrices", J. Math. and Phys. 12, 311-320.

Rankin, R. A.

1964 "Difference Sets", Acta Arith. 9, 161-168.

Reuschle, C. G.

1875 Tafeln Complexer Primzahlen, Welche aus Wurzeln der Einheit Gebildet Sind, Berlin.

Roth, R.

1964 "Collineation Groups of Finite Projective Planes", Math. Z. 83, 409-421.

Ryser, H. J.

1950 "A Note on a Combinatorial Problem", Proc. Amer. Math. Soc. 1, 422-424.

1952 "Matrices with Integer Elements in Combinatorial Investigations", Amer. J. Math. 74, 769-773.

1963 Combinatorial Mathematics, Carus Math. Mono. 14, Wiley, New York.

Schutzenberger, M. P.

1949 "A Non-Existence Theorem for an Infinite Family of Symmetrical Block Designs", Ann. Eugenics 14, 286-287.

Shrikhande, S. S.

1950 "The Impossibility of Certain Symmetrical Balanced Incomplete Block Designs", Ann. Math. Statist. 21, 106-111.

Singer, J.

1938 "A Theorem in Finite Projective Geometry and Some Applications to Number Theory", Trans. Amer. Math. Soc. 43, 377-385.

Skolem, Th., S. Chowla and D. J. Lewis

1959 "The Diophantine Equation $2^{n+2} - 7 = x^2$ and Related Problems", Proc. Amer. Math. Soc. 10, 663-669.

Spence, E.

1967 "A New Class of Hadamard Matrices", Glasgow Math. J. 8, 59-62.

Stanton, R. G., and D. A. Sprott

1958 "A Family of Difference Sets", Canad. J. Math. 10, 73-77.

Storer, J., and R. Turyn

1961 "On Binary Sequences", Proc. Amer. Math. Soc. 12, 394-399.

Storer, T.

1967A Cyclotomy and Difference Sets, Markham, Chicago

1967B "Cyclotomies and Difference Sets Modulo a Product of Two Distinct Odd Primes", Michigan Math. J. 14, 117-127.

Turyn, R.

1960 "Optimum Codes Study", Final Report, Sylvania Electric Products, Inc.

1961 "Finite Binary Sequences", Final Report, (Chapter VI), Sylvania Electric Products, Inc.

Turyn, R.

1964 "The Multiplier Theorem for Difference Sets", Canad. J. Math. 16, 386-388.

1965 "Character Sums and Difference Sets", Pacific J. Math. 15, 319-346.

1968 "Sequences with Small Correlation", in Error Correcting Codes, H. B. Mann, Editor, Wiley, New York, 195-228.

1970 "An Infinite Class of Williamson Matrices", to appear.

Van Der Waerden, B. L.

1949 Modern Algebra vol. 1, Ungar, New York.

Wallis, J.

1969A "A Class of Hadamard Matrices", J. Combinatorial Theory 6, 41-44.

1969B "A Note of a Class of Hadamard Matrices", J. Combinatorial Theory 6, 222-223.

1970 "Combinatorial Matrices", Dissertation, La Trobe University.

Watson, G. L.

1960 Integral Quadratic Forms, Cambridge University Press, Cambridge.

Whiteman, A. L.

1957 "The Cyclotomic Numbers of Order Sixteen", Trans. Amer. Math. Soc. 86, 401-413.

1960A "The Cyclotomic Numbers of Order Twelve", Acta Arith. 6, 53-76.

1960B "The Cyclotomic Numbers of Order Ten", Proc. Sym. in Appl. Math. 10, Amer. Math. Soc., 95-111.

1962 "A Family of Difference Sets", Illinois J. Math. 6, 107-121.

1970 "An Infinite Family of Hadamard Matrices of Williamson Type", to appear.

Yamamoto, K.

1963 "Decomposition Fields of Difference Sets", Pacific J. Math. 13, 337-352.

1967 "On Jacobi Sums and Difference Sets", J. Combinatorial Theory 3, 146-181.

1969 "On the Application of Half-Norms to Cyclic Difference Sets", in Combinatorial Mathematics and its Applications, R. C. Bose and T. A. Dowling, Editors, The University of North Carolina Press, Chapel Hill.

Yates, D. L.

1967 "Another Proof of a Theorem on Difference Sets", Proc. Cambridge Philos. Soc. 63, 595-596.

Lecture Notes in Mathematics

Bisher erschienen/Already published

Vol. 1: J. Wermer, Seminar über Funktionen-Algebren. IV, 30 Seiten. 1964. DM 3,80 / $ 1.10

Vol. 2: A. Borel, Cohomologie des espaces localement compacts d'après. J. Leray. IV, 93 pages. 1964. DM 9,– / $ 2.60

Vol. 3: J. F. Adams, Stable Homotopy Theory. Third edition. IV, 78 pages. 1969. DM 8,– / $ 2.20

Vol. 4: M. Arkowitz and C. R. Curjel, Groups of Homotopy Classes. 2nd. revised edition. IV, 36 pages. 1967. DM 4,80 / $ 1.40

Vol. 5: J.-P. Serre, Cohomologie Galoisienne. Troisième édition. VIII, 214 pages. 1965. DM 18,– / $ 5.00

Vol. 6: H. Hermes, Term Logic with Choise Operator. III, 55 pages. 1970. DM 6,– / $ 1.70

Vol. 7: Ph. Tondeur, Introduction to Lie Groups and Transformation Groups. Second edition. VIII, 176 pages. 1969. DM 14,– / $ 3.80

Vol. 8: G. Fichera, Linear Elliptic Differential Systems and Eigenvalue Problems. IV, 176 pages. 1965. DM 13,50 / $ 3.80

Vol. 9: P. L. Ivănescu, Pseudo-Boolean Programming and Applications. IV, 50 pages. 1965. DM 4,80 / $ 1.40

Vol. 10: H. Lüneburg, Die Suzukigruppen und ihre Geometrien. VI, 111 Seiten. 1965. DM 8,– / $ 2.20

Vol. 11: J.-P. Serre, Algèbre Locale. Multiplicités. Rédigé par P. Gabriel. Seconde édition. VIII, 192 pages. 1965. DM 12,– / $ 3.30

Vol. 12: A. Dold, Halbexakte Homotopiefunktoren. II, 157 Seiten. 1966. DM 12,– / $ 3.30

Vol. 13: E. Thomas, Seminar on Fiber Spaces. IV, 45 pages. 1966. DM 4,80 / $ 1.40

Vol. 14: H. Werner, Vorlesung über Approximationstheorie. IV, 184 Seiten und 12 Seiten Anhang. 1966. DM 14,– / $ 3.90

Vol. 15: F. Oort, Commutative Group Schemes. VI, 133 pages. 1966. DM 9,80 / $ 2.70

Vol. 16: J. Pfanzagl and W. Pierlo, Compact Systems of Sets. IV, 48 pages. 1966. DM 5,80 / $ 1.60

Vol. 17: C. Müller, Spherical Harmonics. IV, 46 pages. 1966. DM 5,– / $ 1.40

Vol. 18: H.-B. Brinkmann und D. Puppe, Kategorien und Funktoren. XII, 107 Seiten, 1966. DM 8,– / $ 2.20

Vol. 19: G. Stolzenberg, Volumes, Limits and Extensions of Analytic Varieties. IV, 45 pages. 1966. DM 5,40 / $ 1.50

Vol. 20: R. Hartshorne, Residues and Duality. VIII, 423 pages. 1966. DM 20,– / $ 5.50

Vol. 21: Seminar on Complex Multiplication. By A. Borel, S. Chowla, C. S. Herz, K. Iwasawa, J.-P. Serre. IV, 102 pages. 1966. DM 8,– / $ 2.20

Vol. 22: H. Bauer, Harmonische Räume und ihre Potentialtheorie. IV, 175 Seiten. 1966. DM 14,– / $ 3.90

Vol. 23: P. L. Ivănescu and S. Rudeanu, Pseudo-Boolean Methods for Bivalent Programming. 120 pages. 1966. DM 10,– / $ 2.80

Vol. 24: J. Lambek, Completions of Categories. IV, 69 pages. 1966. DM 6,80 / $ 1.90

Vol. 25: R. Narasimhan, Introduction to the Theory of Analytic Spaces. IV, 143 pages. 1966. DM 10,– / $ 2.80

Vol. 26: P.-A. Meyer, Processus de Markov. IV, 190 pages. 1967. DM 15,– / $ 4.20

Vol. 27: H. P. Künzi und S. T. Tan, Lineare Optimierung großer Systeme. VI, 121 Seiten. 1966. DM 12,– / $ 3.30

Vol. 28: P. E. Conner and E. E. Floyd, The Relation of Cobordism to K-Theories. VIII, 112 pages. 1966. DM 9,80 / $ 2.70

Vol. 29: K. Chandrasekharan, Einführung in die Analytische Zahlentheorie. VI, 199 Seiten. 1966. DM 16,80 / $ 4.70

Vol. 30: A. Frölicher and W. Bucher, Calculus in Vector Spaces without Norm. X, 146 pages. 1966. DM 12,– / $ 3.30

Vol. 31: Symposium on Probability Methods in Analysis. Chairman. D. A. Kappos. IV, 329 pages. 1967. DM 20,– / $ 5.50

Vol. 32: M. André, Méthode Simpliciale en Algèbre Homologique et Algèbre Commutative. IV, 122 pages. 1967. DM 12,– / $ 3.30

Vol. 33: G. I. Targonski, Seminar on Functional Operators and Equations. IV, 110 pages. 1967. DM 10,– / $ 2.80

Vol. 34: G. E. Bredon, Equivariant Cohomology Theories. VI, 64 pages. 1967. DM 6,80 / $ 1.90

Vol. 35: N. P. Bhatia and G. P. Szegö, Dynamical Systems. Stability Theory and Applications. VI, 416 pages. 1967. DM 24,– / $ 6.60

Vol. 36: A. Borel, Topics in the Homology Theory of Fibre Bundles. VI, 95 pages. 1967. DM 9,– / $ 2.50

Vol. 37: R. B. Jensen, Modelle der Mengenlehre. X, 176 Seite DM 14,– / $ 3.90

Vol. 38: R. Berger, R. Kiehl, E. Kunz und H.-J. Nastold, Diff rechnung in der analytischen Geometrie IV, 134 Seiten. 1967 DM 12,– / $ 3.30

Vol. 39: Séminaire de Probabilités I. II, 189 pages. 1967. DM 14,–

Vol. 40: J. Tits, Tabellen zu den einfachen Lie Gruppen und i stellungen. VI, 53 Seiten. 1967. DM 6.80 / $ 1.90

Vol. 41: A. Grothendieck, Local Cohomology. VI, 106 page DM 10,– / $ 2.80

Vol. 42: J. F. Berglund and K. H. Hofmann, Compact Semito Semigroups and Weakly Almost Periodic Functions. VI, 16 1967. DM 12,– / $ 3.30

Vol. 43: D. G. Quillen, Homotopical Algebra. VI, 157 page DM 14,– / $ 3.90

Vol. 44: K. Urbanik, Lectures on Prediction Theory. IV, 50 pag DM 5,80 / $ 1.60

Vol. 45: A. Wilansky, Topics in Functional Analysis. VI, 102 pag DM 9,60 / $ 2.70

Vol. 46: P. E. Conner, Seminar on Periodic Maps. IV, 116 pag DM 10,60 / $ 3.00

Vol. 47: Reports of the Midwest Category Seminar I. IV, 181 pag DM 14,80 / $ 4.10

Vol. 48: G. de Rham, S. Maumary et M. A. Kervaire, Torsion et Typ d'Homotopie. IV, 101 pages. 1967. DM 9,60 / $ 2.70

Vol. 49: C. Faith, Lectures on Injective Modules and Quotien XVI, 140 pages. 1967. DM 12,80 / $ 3.60

Vol. 50: L. Zalcman, Analytic Capacity and Rational Approxim 155 pages. 1968. DM 13.20 / $ 3.70

Vol. 51: Séminaire de Probabilités II. IV, 199 pages. 1968. DM 14,– / $ 3.90

Vol. 52: D. J. Simms, Lie Groups and Quantum Mechanics. IV, 9 1968. DM 8,– / $ 2.20

Vol. 53: J. Cerf, Sur les difféomorphismes de la sphère de d trois (Γ_4 = O). XII, 133 pages. 1968. DM 12,– / $ 3.30

Vol. 54: G. Shimura, Automorphic Functions and Number Th 69 pages. 1968. DM 8,– / $ 2.20

Vol. 55: D. Gromoll, W. Klingenberg und W. Meyer, Riemanns metrie im Großen. VI, 287 Seiten. 1968. DM 20,– / $ 5.50

Vol. 56: K. Floret und J. Wloka, Einführung in die Theorie der voxen Räume. VIII, 194 Seiten. 1968. DM 16,– / $ 4.40

Vol. 57: F. Hirzebruch und K. H. Mayer, O (n)-Mannigfaltigkeit sche Sphären und Singularitäten. IV, 132 Seiten. 1968. DM 10,8

Vol. 58: Kuramochi Boundaries of Riemann Surfaces. IV, 10 1968. DM 9,60 / $ 2.70

Vol. 59: K. Jänich, Differenzierbare G-Mannigfaltigkeiten. VI, 8 1968. DM 8,– / $ 2.20

Vol. 60: Seminar on Differential Equations and Dynamical Edited by G. S. Jones. VI, 106 pages. 1968. DM 9,60 / $ 2.70

Vol. 61: Reports of the Midwest Category Seminar II. IV, 91 pag DM 9,60 / $ 2.70

Vol. 62: Harish-Chandra, Automorphic Forms on Semisimple Li X, 138 pages. 1968. DM 14,– / $ 3.90

Vol. 63: F. Albrecht, Topics in Control Theory. IV, 65 pag DM 6,80 / $ 1.90

Vol. 64: H. Berens, Interpolationsmethoden zur Behandlung v ximationsprozessen auf Banachräumen. VI, 90 Seiten. 1968. DM 8,– / $ 2.20

Vol. 65: D. Kölzow, Differentiation von Maßen. XII, 102 Seit DM 8,– / $ 2.20

Vol. 66: D. Ferus, Totale Absolutkrümmung in Differentialg und -topologie. VI, 85 Seiten. 1968. DM 8,– / $ 2.20

Vol. 67: F. Kamber and P. Tondeur, Flat Manifolds. IV, 53 pag DM 5,80 / $ 1.60

Vol. 68: N. Boboc et P. Mustată, Espaces harmoniques ass opérateurs différentiels linéaires du second ordre de type e VI, 95 pages. 1968. DM 8,60 / $ 2.40

Vol. 69: Seminar über Potentialtheorie. Herausgegeben von I VI, 180 Seiten. 1968. DM 14,80 / $ 4.10

Vol. 70: Proceedings of the Summer School in Logic. Edited by M IV, 331 pages. 1968. DM 20,– / $ 5.50

Vol. 71: Séminaire Pierre Lelong (Analyse), Année 1967 – 190 pages. 1968. DM 14,– / $ 3.90

Bitte wenden / C

Vol. 144: Seminar on Differential Equations and Dynamical Systems, II. Edited by J. A. Yorke. VIII, 268 pages. 1970. DM 20,– / $ 5.50

Vol. 145: E. J. Dubuc, Kan Extensions in Enriched Category Theory. XVI, 173 pages. 1970. DM 16,– / $ 4.40

Vol. 146: A. B. Altman and S. Kleiman, Introduction to Grothendieck Duality Theory. II, 192 pages. 1970. DM 18,– / $ 5.00

Vol. 147: D. E. Dobbs, Cech Cohomological Dimensions for Commutative Rings. VI, 176 pages. 1970. DM 16,– / $ 4.40

Vol. 148: R. Azencott, Espaces de Poisson des Groupes Localement Compacts. IX, 141 pages. 1970. DM 14,– / $ 3.90

Vol. 149: R. G. Swan and E. G. Evans, K-Theory of Finite Groups and Orders. IV, 237 pages. 1970. DM 20,– / $ 5.50

Vol. 150: Heyer, Dualität lokalkompakter Gruppen. XIII, 372 Seiten. 1970. DM 20,– / $ 5.50

Vol. 151: M. Demazure et A. Grothendieck, Schémas en Groupes I. (SGA 3). XV, 562 pages. 1970. DM 24,– / $ 6.60

Vol. 152: M. Demazure et A. Grothendieck, Schémas en Groupes II. (SGA 3). IX, 654 pages. 1970. DM 24,– / $ 6.60

Vol. 153: M. Demazure et A. Grothendieck, Schémas en Groupes III. (SGA 3). VIII, 529 pages. 1970. DM 24,– / $ 6.60

Vol. 154: A. Lascoux et M. Berger, Variétés Kähleriennes Compactes. VII, 83 pages. 1970. DM 8,– / $ 2.20

Vol. 155: Several Complex Variables I, Maryland 1970. Edited by J. Horváth. IV, 214 pages. 1970. DM 18,– / $ 5.00

Vol. 156: R. Hartshorne, Ample Subvarieties of Algebraic Varieties. XIV, 256 pages. 1970. DM 20,– / $ 5.50

Vol. 157: T. tom Dieck, K. H. Kamps und D. Puppe, Homotopietheorie. VI, 265 Seiten. 1970. DM 20,– / $ 5.50

Vol. 158: T. G. Ostrom, Finite Translation Planes. IV. 112 pages. 1970. DM 10,– / $ 2.80

Vol. 159: R. Ansorge und R. Hass. Konvergenz von Differenzenverfahren für lineare und nichtlineare Anfangswertaufgaben. VIII, 145 Seiten. 1970. DM 14,– / $ 3.90

Vol. 160: L. Sucheston, Constributions to Ergodic Theory and Probability. VII, 277 pages. 1970. DM 20,– / $ 5.50

Vol. 161: J. Stasheff, H-Spaces from a Homotopy Point of View. VI, 95 pages. 1970. DM 10,– / $ 2.80

Vol. 162: Harish-Chandra and van Dijk, Harmonic Analysis on Reductive p-adic Groups. IV, 125 pages. 1970. DM 12,– / $ 3.30

Vol. 163: P. Deligne, Equations Différentielles à Points Singuliers Réguliers. III, 133 pages. 1970. DM 12,– / $ 3.30

Vol. 164: J. P. Ferrier, Seminaire sur les Algebres Complétes. II, 69 pages. 1970. DM 8,– / $ 2.20

Vol. 165: J. M. Cohen, Stable Homotopy. V, 194 pages. 1970. DM 16,– / $ 4.40

Vol. 166: A. J. Silberger, PGL_2 over the p-adics: its Representations, Spherical Functions, and Fourier Analysis. VII, 202 pages. 1970. DM 18,– / $ 5.00

Vol. 167: Lavrentiev, Romanov and Vasiliev, Multidimensional Inverse Problems for Differential Equations. V, 59 pages. 1970. DM 10,– / $ 2.80

Vol. 168: F. P. Peterson, The Steenrod Algebra and its Applications: A conference to Celebrate N. E. Steenrod's Sixtieth Birthday. VII, 317 pages. 1970. DM 22,– / $ 6.10

Vol. 169: M. Raynaud, Anneaux Locaux Henséliens. V, 129 pages. 1970. DM 12,– / $ 3.30

Vol. 170: Lectures in Modern Analysis and Applications III. Edited by C. T. Taam. VI, 213 pages. 1970. DM 18,– / $ 5.00.

Vol. 171: Set-Valued Mappings, Selections and Topological Properties of 2^X. Edited by W. M. Fleischman. X, 110 pages. 1970. DM 12,– / $ 3.30

Vol. 172: Y.-T. Siu and G. Trautmann, Gap-Sheaves and Extension of Coherent Analytic Subsheaves. V, 172 pages. 1971. DM 16,– / $ 4.40

Vol. 173: J. N. Mordeson and B. Vinograde, Structure of Arbitrary Purely Inseparable Extension Fields. IV, 138 pages. 1970. DM 14,– / $ 3.90.

Vol. 174: B. Iversen, Linear Determinants with Applications to the Picard Scheme of a Family of Algebraic Curves. VI, 69 pages. 1970. DM 8,– / $ 2.20.

Vol. 175: M. Brelot, On Topologies and Boundaries in Potential Theory. VI, 176 pages. 1971. DM 18,– / $ 5.00

Vol. 176: H. Popp, Fundamentalgruppen algebraischer Mann keiten. IV, 154 Seiten. 1970. DM 16,– / $ 4.40

Vol. 177: J. Lambek, Torsion Theories, Additive Semantics an of Quotients. VI, 94 pages. 1971. DM 10,– / $ 2.80

Vol. 178: Th. Bröcker und T. tom Dieck, Kobordismenthe 191 Seiten. 1970. DM 18,– / $ 5.00

Vol. 179: Seminaire Bourbaki – vol. 1968/69. Exposés 347- 295 pages. 1971. DM 22,– / $ 6.10

Vol. 180: Séminaire Bourbaki – vol. 1969/70. Exposés 364 310 pages. 1971. DM 22,– / $ 6.10

Vol. 181: F. DeMeyer and E. Ingraham, Separable Algeb Commutative Rings. V, 157 pages. 1971. DM 16.– / $ 4.40

Vol. 182: L. D. Baumert. Cyclic Difference Sets. VI, 166 pag DM 16,– / $ 4.40

ture Notes in Physics

'schienen/Already published

C. Erdmann, Wärmeleitung in Kristallen, theoretische Grund-
d fortgeschrittene experimentelle Methoden. II, 283 Seiten.
20,– / $ 5.50

. Hepp, Théorie de la renormalisation. III, 215 pages. 1969.
/ $ 5.00

Martin, Scattering Theory: Unitarity, Analytic and Crossing.
ages. 1969. DM 14,– / $ 3.90

. Ludwig, Deutung des Begriffs physikalische Theorie und
che Grundlegung der Hilbertraumstruktur der Quantenme-
rch Hauptsätze des Messens. XI, 469 Seiten. 1970. DM 28,– /

1. Schaaf, The Reduction of the Product of Two Irreducible
Representations of the Proper Orthochronous Quantumme-
Poincaré Group. IV, 120 pages. 1970. DM 14,– / $ 3.90

Group Representations in Mathematics and Physics. Edited
gmann. V, 340 pages. 1970. DM 24,– / $ 6.60